Tolley's Risk Assessment Workbook Series
RETAIL

by

Jacqui Welham FIOSH RSP FIIRSM
Total Control (Anglia) Limited

Series Editor: Allan St John Holt FIOSH RSP
Head of Safety, Royal Mail Group

LexisNexis™ UK

Members of the LexisNexis Group worldwide

United Kingdom	LexisNexis UK, a Division of Reed Elsevier (UK) Ltd, 2 Addiscombe Road, CROYDON CR9 5AF
Argentina	LexisNexis Argentina, BUENOS AIRES
Australia	LexisNexis Butterworths, CHATSWOOD, New South Wales
Austria	LexisNexis Verlag ARD Orac GmbH & Co KG, VIENNA
Canada	LexisNexis Butterworths, MARKHAM, Ontario
Chile	LexisNexis Chile Ltda, SANTIAGO DE CHILE
Czech Republic	Nakladatelství Orac sro, PRAGUE
France	Editions du Juris-Classeur SA, PARIS
Germany	LexisNexis Deutschland GmbH, FRANKFURT and MUNSTER
Hong Kong	LexisNexis Butterworths, HONG KONG
Hungary	HVG-Orac, BUDAPEST
India	LexisNexis Butterworths, NEW DELHI
Ireland	Butterworths (Ireland) Ltd, DUBLIN
Italy	Giuffrè Editore, MILAN
Malaysia	Malayan Law Journal Sdn Bhd, KUALA LUMPUR
New Zealand	LexisNexis Butterworths, WELLINGTON
Poland	Wydawnictwo Prawnicze LexisNexis, WARSAW
Singapore	LexisNexis Butterworths, SINGAPORE
South Africa	LexisNexis Butterworths, Durban
Switzerland	Stämpfli Verlag AG, BERNE
USA	LexisNexis, DAYTON, Ohio

First published in 2003

A CIP Catalogue record for this book is available from the British Library.

ISBN 0 7545 1890 6

Typeset by Letterpart Limited, Reigate, Surrey

Printed and bound in Great Britain by Antony Rowe Ltd, Chippenham, Wilts

Visit LexisNexis UK at www.lexisnexis.co.uk

Preface

Risk assessment is carried out in order to evaluate the extent and significance of workplace hazards so that appropriate control measures can be put in place. Usually, a written record of the assessment is required. The busy manager who has to start with a blank sheet of paper faces an uphill struggle – most of us have been in that position at some time. Failure to carry out risk assessments (an issue increasingly coming before the courts) is often due not to lack of will but to lack of a suitable system. This series of Workbooks aims to provide the system, and to provide templates with some of the work already done.

There are many ways of assessing risks, some difficult and some easy to follow. There is now a trend towards task-based assessments, which can be more difficult to carry out. Whichever method is chosen, it is important that the assessments are tailored to the particular circumstances in the workplace so as to evaluate real risks and not theoretical ones. The work done by the authors of the Workbooks is therefore the basis for an assessment system, which needs to be completed in detail by the user.

The Workbooks offer an assessment method that ensures compliance with the law, and a practical system that can be understood easily, put into place quickly where necessary and offers real benefits to the users especially by encouraging updating and review. The rules are fully explained, although the principle is simple. Perhaps the simplest and best summary of what needs to be done was published in the Health and Safety Executive's leaflet 'Five Steps to Risk Assessment' in 1994. Any employer who adheres to these guidelines will comply with the law, and perhaps more importantly, create a safer workplace.

Allan St John Holt

Acknowledgements

The Publishers would like to kindly thank Sypol for allowing their generic risk assessment templates to be adapted for use in the Workbooks.

About the Author

Jacqui Welham

Jacqui Welham FIOSH RSP FIIRSM holds a Diploma from Loughborough University in Occupational Health and Safety Management, and recently completed an LLM Degree (Master of Laws) researching the transposition of EC Directives into Member States' legislation.

Jacqui is Managing Director of Total Control (Anglia) Limited, providing a health and safety consultancy service to clients in a wide variety of organisations, including the retail industry. Risk assessments are undertaken for clients on a regular basis, although the company philosophy is to train employees to undertake risk assessments in their own workplace, where they will have an intimate knowledge of the day-to-day activities of their company. Total Control is a City and Guilds registered NVQ centre for Occupational Health and Safety Practice as well as providing IOSH and CIEH training courses.

Previous books include: *Recreational Diving Health & Safety Handbook*, *Frogman Spy* and *Successful Holidays – and What to do if They are Not.*

Series Editor

Allan St John Holt

Allan St John Holt is one of Britain's best-known safety experts. He has also lectured widely in the USA, the Far East and Australia on health and safety matters. Twice President of the Institution of Occupational Safety and Health (IOSH), and a Fellow of the Institution, he is a Registered Safety Practitioner. He is currently Head of Safety for the Royal Mail Group.

A social studies graduate, Allan founded the National Examination Board in Occupational Safety and Health in 1979. An Ambassador and member of the Board of Directors of Veterans of Safety International, he was inducted into the Safety and Health Hall of Fame International in October 1997, and was elected Honorary President of the Southampton Occupational Safety Association in 1991.

Contents

Part 1

Part 2

Part I

Risk Assessment

Introduction

Risk assessment is not a new idea. The principle of relating the amount of action required to the size of the problem was set out almost 30 years ago within the *Health and Safety at Work etc Act 1974*. This was the first major piece of health and safety law to avoid giving directions and rely instead upon a management evaluation of what is needed to take reasonable care of the health and safety of both employees and anyone else affected by the work activities of an organisation. The employer could only do what is 'reasonably practicable' by evaluating the hazards and risks involved and applying appropriate controls.

The first health and safety law to require a specific evaluation, or risk assessment, was the *Control of Lead at Work Regulations 1980 (SI 1980 No 1248)*. This was followed seven years later by Regulations controlling work with asbestos (*Control of Asbestos at Work Regulations 1987 (SI 1987 No 2115)*) and then by the *Control of Substances Hazardous to Health Regulations 1988 (SI 1988 No 1657) (COSHH Regulations)* which came into force in October 1989. In 1993 the set of Regulations based upon new European Council Directives on health and safety was based on the principle of risk assessment. In many ways, the most important of these was the *Management of Health and Safety at Work Regulations 1999 (1999 No 3242) (MHSWR 1999)*, introducing risk assessment as a requirement for all work situations other than those already covered by more specific assessment requirements.

The Regulations, although now revised and expanded from their original form, do not specify any particular method of carrying out risk assessments and, as a result, there are many different systems in use. Experience shows that the process does not have to be especially time consuming provided that a format is used that presents and records information in a clear and logical way. One of the few direct requirements of all the Regulations on the subject is that the results must be recorded in writing where there are five or more employees, so the employer's chosen format is important in order to show that his/her assessments are suitable and sufficient.

Risk assessment is the key to successful management of health and safety at work. The law no longer provides a list of 'what to do': it now recognises that life is more complicated than that. Employers must work out for themselves what the problems are:

- the hazards – and how much of a problem each one is;
- the risk – and then work out appropriate control measures.

This is the essence of risk assessment. It is a continuous process, and a full explanation of it in the context of the systematic management of health and safety can be found in the Health & Safety Executive's (HSE) publication HSG65 'Successful Health and Safety Management'.

Once the assessment is completed and recorded, the document becomes a blueprint for action to remove hazards and lower the risks associated with each work activity. Only then can the employer establish policies to control the risks he is now aware of, and set up procedures. Risk assessments also need to be reviewed, as conditions change over time. *Management of Health and Safety at Work Regulations 1999 (1999 No 3242) (MHSWR 1999)* and specific legislation specifies the need to review; this must be done:

- when new materials and equipment are introduced;
- following significant changes in work practices and premises, and when monitoring shows that control measures are not working properly.

Following accidents, the assessments should be reviewed for accuracy, to make sure that the circumstances have been taken fully into account.

The key is to identify significant hazards and risks. In some cases, where the situation changes rapidly because of the nature of the work, risk assessment will identify a range of hazards and their attendant risks and specify general controls, relying upon selection of employees and appropriate training to enable them to take control of specific risks as they arise.

Risk assessments can be made of premises, machines, processes and tasks. In some circumstances, it has been found easiest to make assessments that are task or process based – common examples include driving company vehicles, using fork lift trucks, loading and unloading goods.

Terminology

Some definitions

The words 'hazard' and 'risk' have special meanings in health and safety, which are narrower and more specific then their customary definitions.

Hazard means the inherent property or ability of something or some situation to cause harm, loss or damage. A hazard has the potential to interrupt or interfere with a process or a person. **Risk** is the chance, likelihood or probability of the hazard actually causing the loss – an evaluation of the potential within the hazard.

So a simple way to remember the difference is that 'hazard' describes potential for harm, and 'risk' is the chance that harm will result in the particular circumstances.

Risk has two main components, the likelihood of the failure occurring, and the severity of the consequences of failure. Comparisons between risks can be made by giving numerical values to these factors. This is helpful in allowing an estimate to be made of the reduction in the risk that can be made by the introduction of controls, and also in establishing priorities for action.

It is important to be aware of the limitations in the use of numerical risk ratings. For example, the probability of failure and the extent of the outcome cannot be predicted for individuals in many cases of occupational ill-health exposure. As a result, numerical scoring of risk is unreliable for evaluating the extent of occupational health hazards. Some alternative methods of carrying out and recording risk assessments use a simple classification such as Insignificant, Low, Medium and High in order to avoid any confusion or argument about what numbers might mean in any particular circumstance.

Risk assessments are required to be 'suitable' and 'sufficient' for the purpose. Such an assessment will reflect what it is reasonably practicable for the assessor to know or find out about the hazards in the workplace. Therefore, the assessor must be **competent**, and this means someone with an appropriate combination of knowledge, training and experience in the industry to enable them to:

- evaluate the impact of specific legal requirements and the need for risk assessments;
- formulate an appropriate action plan to remove the hazards and control the risks;
- co-ordinate the implementation work.

Another part of *MHSWR 1999* requires management to appoint one or more competent persons, normally from amongst the employees, to assist the employer in complying with health and safety law. This person should either carry out or oversee the risk assessment process. Ability to do so will be one of the key competencies of the person so appointed.

(The section on RISK ASSESSMENT IN THE RETAIL INDUSTRY gives details of specific Regulations applying to Manufacturing.)

Control measures

Some of the controls that can be used work better than others, and some can be very ineffective. In order of effectiveness, these are the accepted headings of controls available:

- **Hazard elimination –** the best solution of all is to remove the hazard, use of alternatives, design improvements, or just deciding not to attempt the work.
- **Substitution** – replacing a hazard can lower the risk, such as when replacing one chemical with another less dangerous one, or using a more appropriate (safer) piece of equipment.

- **Use of barriers** – **isolation**, such as putting the hazard in a box, for example the physical isolation of noisy equipment, and **segregation** (removing employees from the hazard by putting them in a box such as a protected cabin).
- **Use of procedures** – procedures can limit exposure time, safe systems of work can define acceptable limitations on the work – but these depend upon people to carry them out properly and they are less effective as a result – they can be easily avoided or ignored.
- **Use of warning systems** – signs, instructions, labels – all of which depend on human response and again can be ignored.
- **Use of PPE** – this depends totally on human response to be effective, assuming an appropriate piece of equipment has been selected. This is to be used as a sole measure only when all other options have been exhausted – personal protective equipment (PPE) is the last resort.

Putting it all together

Hazards and risks are sometimes difficult to identify and separate, so it will be helpful at this stage to think about a common activity that everyone can relate to and then use the Workbook format to analyse and record the assessment.

Most businesses use motor vehicles and so the activity of driving vehicles at work needs to be assessed. First, we can list the main or significant **hazards** associated with motoring:

- collision with other vehicles and roadway structures;
- collision with permanent structures in the premises;
- collision with pedestrians in the premises;
- objects falling from the vehicle;
- vehicle defects;
- fire;

and there may be others, such as carriage of dangerous goods, specific to the business.

The **risks** associated with these hazards will vary considerably with the particular circumstances, but some general statements can be made about them that will be valid and useful. For example, fire is relatively uncommon, and few people are involved in vehicle fires. On the whole, we can say that the fire risk is low, but the severity can be high. If flammable substances are carried in the vehicle the risk level would rise, and that shows **why risk assessments must always be related to the particular circumstances of the work and not just made for the general case**.

Other factors affecting risk will vary according to the individual circumstances. We have no influence on most of them, for example:

- stopping distance (speed and size, condition of vehicle, road conditions);
- speed and size of other vehicles;
- skill of drivers – which is related to age, experience, reaction times, physical condition;
- provision of safety devices for occupants;
- environmental conditions;
- traffic conditions;
- evasive skills of pedestrians.

What **controls** are already in place? A convenient classification for them is:

- Statutory – regulations concerning construction and use, licensing of drivers, requiring vehicle testing and speed restrictions.
- Guidance – Highway Code and other information resources.
- Planning – daily and weekly checks on the vehicle, checking the route before starting out and checking on the condition of the driver.
- Physical – measures include pedestrian crossing points, roadside barriers, signposts, seatbelts.
- Monitoring – carried out by the police, roadside cameras, spot checks on vehicles, and the monitoring generally of accidents and traffic patterns.

- Training – driver training is controlled and standards are laid down. Training of vulnerable groups of pedestrians (such as school children) is done using the Highway Code but is not a legal requirement.

Therefore, all these measures are already in place. The task of the risk assessment is to look beyond the general to the particular. This will mean assembling information, especially previous incidents. The list of hazards can be revisited and related specifically to the work being done, and can be set out as follows.

Basic risk assessment

Hazard	Risk level	Who is at risk?	Additional controls needed	Action by:
Collision with other vehicles and roadway structures	Low – low vehicle mileage, no significant accident record	Drivers, third parties	None	
Collision with permanent structures in the premises	Low – some structural damage recorded	Drivers	Improved road markings, road signs	Within 28 days
Collision with pedestrians in the premises	High	Third parties	Restrict access, separate pedestrians from roads	Within 7 days
Objects falling from the vehicle	Low – no unusual or unsecured loads carried	Third parties	None	
Vehicle defects	Low	Drivers	Planned maintenance system to be reviewed	Within 28 days
Fire	Medium – occasional flammable loads carried short distances within the premises	Drivers	Supply vehicle fire extinguishers to identified vehicles	Within 7 days

Rating the risks

Quantitative risk assessments produce a probability estimate based upon known risk information applied to the circumstances being considered. They can be found in 'safety cases' in high risk industries where, for example, they might indicate the probability of a nuclear incident or serious chemical leak.

Qualitative assessments are subjective, based on the personal judgement of the assessor, backed by more general information about probability. These are much simpler to make, and are those normally referred to by legal requirements. We know this because the HSE published guidance in 1994 entitled 'Five Steps to Risk Assessment', a leaflet setting out the most basic steps in the process and providing a specimen format for recording the findings of assessments. The leaflet has since been updated and reissued. Comments contained in the text give useful pointers to the approach to risk assessment which will satisfy minimum requirements in the view of the enforcing authority. For example:

> 'An assessment of risk is nothing more than a careful examination of what, in your work, could cause harm to people so that you can weigh up whether you have taken enough precautions or should do more . . .'.

We are told there is no need to make assessments complicated. Nevertheless, there are benefits to be gained by estimating the level of risk associated with hazards, in particular because we can then make an evaluation of the expected benefits of the control measures contained in the assessment – a before and after evaluation. The choice of whether to do this numerically, or descriptively, or at all, is left to the individual.

A numerical approach has its drawbacks. Two or more variables can be used to produce a final rating by multiplication, as shown in the following Table. But there should be a method of ensuring that any high-scoring severity evaluation forces the problem to be studied at an early stage – even if the multiplication exercise puts it low in the priority list. For example, an event with a high probability of occurrence but a low severity rate could be rated the same as an event with a low probability of occurrence but a high severity rate. The numbers themselves have no inherent value – a risk rated at 20 is not 'half as dangerous' as one rated at 10. The purpose of the exercise is simply to allow priorities to be established and to demonstrate the effectiveness of the controls put in place.

A descriptive approach simply uses words such as 'insignificant', 'low', 'medium' and 'high' in their normal sense to evaluate the product of the likely severity of the outcome and the probability of its occurrence. Most people have an appreciation of what 'Medium Risk' means for them in their business context, and can appreciate that with controls in place the risk should be lower – preferable to 'Insignificant' or at the least to 'Low'.

Numerical Assessment Table

Probability (likelihood of the event)	
Almost certain	5
Probable	4
Could happen occasionally	3
Improbable	2
Almost impossible to happen	1

Severity (the most likely result of the accident)	
Catastrophic – death	5
Severe incapacity	4
Absent one month	3
Absent three days + recovery	2
First aid treatment required only	1

Assess the risk, multiply the two numbers and transfer to the **Risk Index Table**:

		Probability				
Severity		5	4	3	2	1
	5	25	20	15	10	5
	4	20	16	12	8	4
	3	15	12	9	6	3
	2	10	8	6	4	2
	1	5	4	3	2	1

Typical actions required

Priority 1	25–16	Extensive precautions based on physical controls, work method statements, specific and stringent rules with significant penalties for infringement.
Priority 2	15–10	Training, checks on competence of employees needed, high level of supervision.
Priority 3	9–1	Some risk acceptable, competent employees appropriate level of supervision.

Overview of the law regarding risk assessment

Health and Safety at Work etc Act 1974

This Act is the central piece of health and safety legislation in Great Britain. The bulk of the Act is now historical, as it 'enabled' the mechanisms for establishing the Health and Safety Commission and Executive, the Medical Advisory Service (EMAS) and so forth. The substantive portions are contained in the General Duties sections of the *Health and Safety at Work etc Act 1974 (HSWA 1974), ss 2–9.*

General duties of employers are set out in *sections 2–4*, and *9*. As far as is reasonably practicable, employers must safeguard the health, safety and welfare of their employees and anyone else who may be affected by the work activities. This duty extends to providing and maintaining:

- safe plant and systems of work;
- safe handling, storage, maintenance and transport of work articles and substances;
- necessary information, instruction, training and supervision;
- a safe place of work and means of getting to and from it safely;
- a safe working environment with adequate welfare facilities.

The details are left to the employer to work out, but by 1992 a set of six (The Six Pack) more specific pieces of safety law was added to cover both general issues (management, safe workplace and equipment) and specific ones such as personal protective equipment and manual handling. The requirement to make risk assessments was not a feature of the Act, but it can be seen that in order to discharge the general duties, it will be necessary for employers to make an evaluation of health and safety issues facing them. The specific risk assessment requirement for all employers is contained in the *Management of Health and Safety at Work Regulations 1999 (SI 1999 No 3242)*, which first appeared in 1992.

Management of Health and Safety at Work Regulations 1999

The *Management of Health and Safety at Work Regulations 1999 (SI 1999 No 3242) (MHSWR 1999)* provide most of the detail lacking in the *Health and Safety at Work etc Act 1974*, and importantly specify the requirements for risk assessment of all work activities. It is strongly advised to consult a copy of the accompanying Approved Code of Practice (ACoP) which contains both the Regulations and an authoritative interpretation. Employers should be familiar with all of the Regulations, but in relation to risk assessment the following are particularly relevant.

Regulation 3

Under *MHSWR 1999 (SI 1999 No 3242), reg 3*, every employer and self-employed person is to make a suitable and sufficient assessment of the health and safety risks to his employees, and others not in his employment to which his undertakings give rise, in order to put in place appropriate control measures. It requires appropriate review to be made of the assessments, and for the significant findings to be recorded if five or more are employed. Details are also to be recorded of any group of employees identified by an assessment as being especially at risk.

Regulation 3 also requires employers to make a specific risk assessment where 'young persons' are employed (these are people under the age of 18). In making or reviewing the assessments, the following must be taken into account:

- the inexperience, immaturity and lack of awareness of risks of young people;
- the fitting-out and layout the of the workplace;
- nature, degree and duration of exposure to physical, chemical and biological agents;
- type, range and use of work equipment;
- organisation of processes and activities;
- risks from special processes which are listed.

A similar provision covering specific risk assessment of new and expectant mothers is to be found in **Regulation 16**.

Regulation 5

MHSWR 1999 (SI 1999 No 3242), reg 5 requires the employer to make appropriate arrangements, given the nature and size of his operations, for effective planning, organising, control, monitoring and review of the preventive and protective measures he puts in place. If there are more than five employees the arrangements are to be recorded. (The measures themselves will have been derived from the risk assessments.)

Regulation 6

MHSWR 1999 (SI 1999 No 3242), reg 6 requires the employer to provide appropriate health surveillance, identified as being necessary by the risk assessments.

Regulation 7

MHSWR 1999 (SI 1999 No 3242), reg 7 deals with the appointment of adequate numbers of competent persons to assist the employer to comply with obligations under all the health and safety legislation, unless (in the case only of sole traders or partnerships) the employer already has sufficient competence to comply without assistance. *Regulation 7* and the ACoP set out clearly what is required of the competent person and the appointment arrangements.

Regulation 8

MHSWR 1999 (SI 1999 No 3242), reg 8 requires employers to establish and give effect to procedures to be followed in the event of 'serious and imminent danger' to persons working in their undertakings, to nominate competent persons to implement any evacuation procedures and restrict access to danger areas. The identification of potential 'serious and imminent danger' is one of the functions of the risk assessment process.

Regulation 10

MHSWR 1999 (SI 1999 No 3242), reg 10 covers the provision of information by the employer on health and safety risks identified by the assessments, plus all the control measures, together with information on the identity of competent persons appointed under *Regulation 8*. There is also more on the employment of children and young persons – before employing them, the employer must provide the parents with comprehensible and relevant information on assessed risks and the control measures in place.

Specific risk assessment legislation summaries

Control of Substances Hazardous to Health Regulations 2002

First introduced in 1988, the *Control of Substances Hazardous to Health Regulations 2002 (SI 2002 No 2677) (COSHH Regulations)* gave the first general requirement for risk assessment, as almost all employers have, make or use substances falling within scope. Although modified and expanded since the original version, the essence of *COSHH* remains the same: prevention of workplace disease and ill health resulting from exposure to hazardous substances. Again, there is a control framework beginning with the assessment of risks to health arising from work activities associated with hazardous substances, introduction of control measures, maintenance of the measures and associated equipment, and monitoring their effectiveness.

Health and safety law also requires information to be made available by the supplier of substances; this provides a good starting point for carrying out risk assessments. Possession of a data sheet is not the same as having done a risk assessment, though. Under the *COSHH Regulations*, there is a duty on the employer to ensure that exposure to substances hazardous to health is either prevented, or where this is not reasonably practicable, controlled. Prevention must initially be attempted by means other than the provision and use of personal protective equipment by employees.

Any substance used at work, or deriving from work activities, is a 'COSHH substance' if it is harmful to people's health in the form in which it occurs at work. Most common substances meeting this definition will be labelled on the container as 'dangerous', using an orange diamond with a black symbol inside it to indicate the contents are toxic, very toxic, harmful or corrosive. However, substantial quantities of dust are also included in the *COSHH* definition, and there may be substances produced as by-products of the work processes that cannot be labelled in this way.

Personal Protective Equipment at Work Regulations 1992

The *Personal Protective Equipment at Work Regulations 1992 (SI 1992 No 1144), reg 4* places a duty on the employer to provide 'suitable personal protective equipment (PPE) to employees, except where risks have been adequately controlled by other equally, or more effective, means. It is to be used as a last resort, steps having been taken to remove hazards or control risks by means of safer processes, systems or conditions. The need for PPE will have been identified by risk assessment.

Regulation 6 requires the employer to make a specific risk assessment to determine whether the proposed PPE is in fact 'suitable' – appropriate for the risks involved and the conditions, capable of fitting the wearer properly after adjustment, and whether it will be effective in controlling the risk.

The assessment must include an evaluation of the risks that have not been avoided by any other means, and a comparison of the characteristics the PPE needs to have to be effective against the risks against the characteristics of the proposed PPE. For high-risk situations and/or where the PPE is complex, the assessment should be in writing and kept available.

Where PPE is to be used, the employer should be aware of other detailed requirements of these Regulations which relate to training and information, storage and maintenance of the PPE.

Manual Handling Operations Regulations 1992

The *Manual Handling Operations Regulations 1992 (SI 1992 No 2793*) focus attention on ergonomic solutions to manual handling issues, again based upon risk assessment. *Regulation 4* is of crucial importance, requiring employers to avoid the need for employees to carry out those manual handling operations that involve a risk of injury as far as is reasonably practicable. Where this cannot be done, a 'suitable and sufficient' risk assessment is required for all such operations, so as to take the appropriate steps to reduce the risk of injury to the lowest level reasonably practicable. The ACoP to the Regulations is particularly helpful as it contains a specimen assessment form and worked examples.

Assessments are required to cover specific questions, which are contained in the Schedule to the Regulations, and to some extent this prescribes the format of the assessment to be used.

Health and Safety (Display Screen Equipment) Regulations 1992

The *Health and Safety (Display Screen Equipment) Regulations 1992 (SI 1992 No 2792)* are aimed at improving working conditions at display screen equipment by providing ergonomic solutions, as well as to enable users of the equipment to obtain information about the hazards, risks and their control measures together with eye and eyesight tests. The specific risk assessment requirement is found in *Regulation 2*, where employers are required to carry out a suitable and sufficient analysis of each workstation, and to review following any significant changes. Major changes in software used and task requirements, hardware and work equipment would be classed as significant. Unusually, the Regulations contain a list of minimum requirements for workstations in the Schedule to the Regulations; this is derived from the EU Directive (*90/279/EEC*).

The need to comply

A failure to carry out risk assessments of an adequate kind renders the employer vulnerable to criminal prosecution and, in relation to specific requirements of Regulations, including the *Health and Safety (Display Screen Equipment) Regulations 1992 (SI 1992 No 2792)* and the *Manual Handling Operations Regulations 1992 (SI 1992 No 2793)*, to successful civil claims for compensation by injured employees.

The level of penalties applied so far by the courts for failure to make risk assessments is becoming significant. This offence is usually not the sole charge brought against an employer, but it is a charge added in an increasing proportion of cases. Put at its simplest, the employer without any or any adequate, suitable and sufficient risk assessments cannot show that the most basic of tasks required of him has been completed – that of having an effective system for managing health and safety at work. Thus he is vulnerable to charges under the Act, as well as several Regulations summarised here.

The good news is that the process is not normally complicated, or one that requires special training beyond an appreciation of what is being done and a knowledge of the particular business.

This workbook supplies the means for basic risk assessment in manufacturing industry, giving templates and worked examples of common hazards and their usual control measures. Details of applicable legislation can be found in the section on RISK ASSESSMENT IN THE RETAIL INDUSTRY.

Table of legislation

- *Noise at Work Regulations 1989 (SI 1989 No 1790);*
- *Management of Health and Safety at Work Regulations 1999 (SI 1999 No 3242);*
- *Health and Safety (Display Screen Equipment) Regulations 1992 (SI 1992 No 2792);*
- *Workplace (Health, Safety and Welfare) Regulations 1992 (SI 1992 No 3004);*
- *Provision and Use of Work Equipment Regulations 1998 (SI 1998 No 2306);*
- *Personal Protective Equipment at Work Regulations 1992 (SI 1992 No 2966);*
- *Health and Safety (Miscellaneous Amendments) Regulations 2002 (SI 2002 No 2174);*
- *Control of Substances Hazardous to Health Regulations 2002 (SI 2002 No 2677);*
- *Fire Certificates (Special Premises) Regulations 1976 (SI 1976 No 2003);*
- *Fire Precautions (Application for a Certificate) Regulations 1989 (SI 1989 No 77);*
- *Fire Precautions (Non-certificated Factory, Office, Shop and Railway Premises) (Revocations) Regulations 1989 (SI 1989 No 78);*
- *Fire Precautions (Workplace) Regulations 1997 (SI 1997 No 1840) (as amended by SI 1999 No 1877);*
- *Fire Safety and Safety of Places of Sport Act 1987.*

Guidance material

Health & Safety Executive:

- L22 – 'Safe Use of Work Equipment – Guidance on Regulations';
- L23 – 'Manual Handling – Guidance on Regulations';
- L25 – 'Personal Protective Equipment at Work – Guidance on Regulations';
- L26 – 'Display Screen Equipment – Guidance on Regulations';
- HSG53 – 'RPE – Selection, Use and Maintenance';
- HSG88 – 'Hand-arm Vibration';
- HSG121 – 'A pain in your workplace – ergonomic problems and solutions';
- HSG218 – 'Tackling work-related stress';
- INDG171 – 'Upper Limb Disorders – Assessing the Risk';
- INDG341 – 'Stress leaflet';
- INDG281 – 'Short guide to stress'.

Home Office:

- Code of Practice: 'Fire Precautions in Factories, Offices, Shops and Railway Premises not required to have a fire certificate';
- 'Fire Precautions Act 1971: Guide to fire precautions in existing places of work that require a fire certificate';
- 'Guide to fire precautions in existing places of entertainment and like places';
- 'Fire safety: an employer's guide', ISBN 0 11 341229 0.

How to use the workbook and methodology

Essential steps in risk assessment

The five basic requirements for successful risk assessment are:

- identify the hazards;
- decide who might be harmed (at risk), and how;
- evaluate the extent of the risk by looking at the action already being taken, then deciding if that is sufficient or if more needs to be done;
- recording the significant findings of the assessment;
- reviewing the assessment at appropriate intervals.

The Workbook provides a format to prompt and help identify the hazards, and to record the actions taken and those needed. Whether part of a review or a first risk assessment for any business, this action plan that will cover all the requirements if followed. The schematic approach summarised below can be used to explain to the assessment team the steps involved and their part in the work.

Action Plan

1. **Appoint one or more competent persons** to co-ordinate the risk assessment activities, with responsibility for checking the adequacy of current policy, arrangements and management systems for ensuring health and safety. The competent person should have access to the Approved Codes of Practice and guidance associated with the Regulations. Their task will be to evaluate the impact of the requirements and especially the need for risk assessments, carry out or supervise the carrying out of the risk assessments, put together a management plan to take the necessary actions and champion the actions. This workbook provides the means of making the required risk assessments and for sharing the results with employees (by training) and with other employers affected by the work activities.
2. **Assemble information.** Safety data sheets supplied by manufacturers of substances used at work, information on substances generated during work, makers' specifications on matters such as maintenance of equipment, plans and designs, workstation layouts, training records – depending upon the activities, all of the above and more may be helpful to the assessment process.
3. **Identify and list all the work activities**. Not all of them will be required to be assessed, if there are no hazards associated with them, or if the risks are insignificant.
4. **Identify all the hazards associated with each activity, substance, procedure and group of people.** It has been found that activity-based assessments are easiest to carry out, but there will be circumstances when individual pieces of equipment and locations will need to be assessed. The Workbook provides examples of these, and covers all the most common workplaces and work practices in the industry.
5. **Carry out risk assessments.** The templates in PART 2 of this Workbook have been specially designed to help with the process. Their use is explained in the section on RISK ASSESSMENT IN THE RETAIL INDUSTRY. The basic principle is that they list the majority of hazards and controls likely to be appropriate in particular situations, and allow their assessment either numerically or descriptively. But use of the Workbook's examples alone is not sufficient: there are likely to be circumstances and hazards that they do not cover, and also the levels of risk associated with each case must be evaluated. Remember that an assessment which is both suitable and sufficient reflects what it is reasonably practicable to expect employers to know about the hazards specific to their workplaces. Evaluate the risk in workplaces by thinking about all relevant factors – including the numbers of people exposed, the duration of exposure, and the likely outcome of the risk if not controlled.
6. **Record basic control measures.** Write down the controls that are already in place, and also what extra needs to be done in order to lower the risk level as far as reasonably practicable. Assign actions and timescales to make sure the actions are carried out.
7. **Record the residual risk.** 'Residual risk' is the level of risk left after the controls have been put in place. Could injury still result? How many are exposed? How likely is it to happen with all the controls in place? To show the suitability and sufficiency of the risk assessment, the effectiveness of the control measures should be evaluated to produce a 'before and after' picture.
8. **Make sure the assessment is dated.** Also, give a date by which the assessment must be re-evaluated. Part of the accident investigation process will be to revisit the assessments to ensure they remain appropriate in the light of any new evidence, but they also need to be checked for adequacy in the light of current work practices and conditions.

Special cases – stress

Stress is the adverse reaction people have to excessive pressure or other types of demand placed upon them. Stress is, therefore, not 'good', or 'an illness'; the Industrial Injuries Board has ruled that stress-related illness at work is not to be classed as an industrial accident. At least one employee in five rate their job as 'very stressful'. Recent work carried out by the HSE shows that management standards for the control of stress need to be reviewed, and the risks assessed. There is a clear duty under the *Health and Safety at Work etc Act 1974* and the *MHSWR 1999* to do so. The same applies to the joined issues of violence and bullying at work.

The liability of the employer to pay compensation to an affected employee depends upon the facts of the particular case. There must always be:

- a duty of care owed to the employee;
- a failure to satisfy that duty;
- an injury or loss as a consequence.

The number of cases claiming compensation for stress-related illness has been increasing rapidly in recent years, but an Appeal Court ruling in February 2002 set out new guidelines for courts after setting aside three awards of £200,000 in total that had been made by lower courts to employees suffering from stress at work.

The Court of Appeal ruled that employees must raise concerns over stress at work with their employer before submitting a claim, in order to have any chance of gaining compensation. This means that employers can normally assume that their employees can withstand the pressures of the job unless they know of some particular problem or vulnerability and, where the employer offers a confidential counselling service with access to treatment, he will rarely be held in breach of the duty of care.

It follows that employees will need to raise their concerns regarding stress with their employers and give the employer the chance to do something about it. If they do not give the employer this opportunity they will find it hard to prove a breach of care. Equally, the employer will also have an obligation to provide the means of raising those concerns. The same arguments are likely to hold good for violence and bullying.

At the time of writing, the HSE's view is that the voluntary route to stress management at work should be explored before any legislation. A European Directive may be produced which would then force specific legislation to be introduced into the UK.

The HSE believes that management must strive to understand the issue, investigate absences from work and analyse them for cause. It advocates that risk assessment within a business should concentrate on seven major stressors (stress factors). These factors are:

1. the demands upon the individual – workload, exposure to hazards;
2. the extent of the control the individual possesses over the work to be done;
3. the level of support, i.e. training, peer support and individual factors;
4. relationships at work – bullying, harassment;
5. nature of the work role – ambiguity, conflicting roles;
6. the pressure for change within the business; and
7. the culture of the business – in relation to workload expectation, for example – is it the norm to work longer hours or take work home?

The following checklist can be used to assess the level of stress risk in a business or part of a business. Numerical values can be given to the answers to produce a score, but the aim should be to produce an action list which will address the deficiencies and thus reduce the risk level. The completed checklist should be dated, signed by the assessor(s), and kept on file.

Checklist

Demands upon the individual	**Yes/ No**	**Required actions**	**Person responsible**	**Completion date**
Does the recruitment process correctly match skills and ability requirements to the job?				
Is the allocation of work monitored?				
Is induction and job change training given and is refresher training available?				
Are the hours to be worked agreed on recruitment and realistic for the work required?				
Do staff understand the reasons for any temporary increases in work or production rates?				

Demands upon the individual	**Yes/ No**	**Required actions**	**Person responsi-ble**	**Comple-tion date**
Is there a monitoring system for working environment factors such as noise, vibration, humidity?				
Are the resources adequate to do the required work?				
Is there suitable and sufficient work equipment available for all jobs?				
Control over the work				
Is a job design process followed for all new jobs, including evaluation?				
Are the Working Time Regulations complied with and work patterns controlled?				
Do employees have an opportunity to plan their own work where possible?				
Is there an element of flexibility in task allocation and hours of work?				
Is there a leadership feedback process?				
Do management and staff meet together regularly to make decisions about the resolution of work issues?				
Support, training and other factors				
Are employees aware of the availability of medical and welfare support teams and how to get in touch with them?				
Is there a clear, open and transparent process for transfers and promotions?				
Is there a process for training and career development, and has it been explained to employees?				
Is there a process for assessing skills and competencies?				
Relationships				
Are staff made aware of the process to deal with harassment?				

Demands upon the individual	**Yes/ No**	**Required actions**	**Person responsi-ble**	**Comple-tion date**
Is there a grievance procedure in place?				
Is there an equal opportunities policy in place, which is complied with?				
Is there measurement of attendance and is an appropriate procedure used correctly?				
Is there a Code of Conduct for the business and is it used properly?				
Is there a system for reporting and controlling bullying, harrassment, drink or drug abuse?				
Role				
Is there an appraisal process, and is it being followed correctly?				
Are there clear job descriptions for each job?				
Are meetings with managers held regularly and the results recorded and shared with employees?				
Does the training process ensure that the correct training is available for all jobs?				
Change				
Are staff consulted and involved in changes that may affect them?				
Is 'change' risk assessed in terms of its likely impact on health and safety?				
Culture				
Is there any stress awareness training?				
Is there a process in place to deal with harassment and bullying?				
Is sick absence monitored regularly for signs of stress related illness?				
Is annual leave monitored to ensure that employees take their leave entitlements?				

Demands upon the individual	Yes/ No	Required actions	Person responsible	Completion date
Are staff discouraged from working excessive hours?				
Are reports of stress taken seriously and action taken?				
Are there good communication channels set up so that employees are kept well informed about the business?				
Is there a feedback process in place for staff comments and/or complaints?				

Special cases – fire

A fire risk assessment must be carried out for every workplace, regardless of the existence of any approval or certification which may have been given. It is a practical exercise, requiring a tour to verify its completeness and accuracy. The purposes of the assessment are to:

- identify the extent of the fire risk;
- assess the likelihood of a fire occurring;
- identify any additional precautions that may be needed, or control systems that do not function adequately.

The nature of the risk assessment is essentially the same – the tasks are:

- identify potential fire hazards;
- decide who might be harmed or at significant risk in the event of a fire, and identify their location;
- evaluate the levels of risk from each hazard, and decide whether existing precautions are adequate – or whether the hazard can be removed or controlled more effectively;
- record the findings and advise employees of the results, prepare or revise emergency plan;
- review the assessment as necessary, as conditions change and at regular intervals in any case.

Particular attention is drawn to the Home Office 'Employer's guide' as a mine of information. The use of a simple code can be convenient to describe the extent of compliance, the condition or the degree of control. The following Table gives an easily-understood illustration.

Code		Definition
A	Fully satisfactory	Meets all requirements, does not require improvement
B	Adequate	Some improvement possible
C	Less than adequate	Significant improvement required, action must be taken
D	Poor	Action is required urgently to improve the condition
N/A	Not applicable	Condition or circumstances not present or applicable

Alternatively the same numerical scoring system can be used as described elsewhere.

Legal summary

Major provisions are currently contained in the *Fire Precautions Act 1971*, as amended by the *Fire Safety and Safety of Places of Sport Act 1987*, and in detailed provisions concerning fire certification. The local fire authority enforces both the Fire Regulations and the parts of the *Management of Health and Safety at Work Regulations 1999 (SI 1999 No 3242)*, which deal with precautions for the safety of people where fires may occur. The latter Regulations apply to the taking of fire precautions in general terms, requiring employers to identify circumstances where situations presenting serious and imminent danger to employees could occur. These must be identified in risk assessments, and procedures written down to control risks. The *Fire Precautions (Workplace) Regulations 1997 (SI 1997 No 1840)* as amended, apply to the taking of fire precautions in premises not required to have a fire certificate.

It is anticipated that the present legal framework will change shortly with a view to simplification, but it is not considered likely that the requirements for fire risk assessment will be altered significantly.

Risk assessment – the action plan

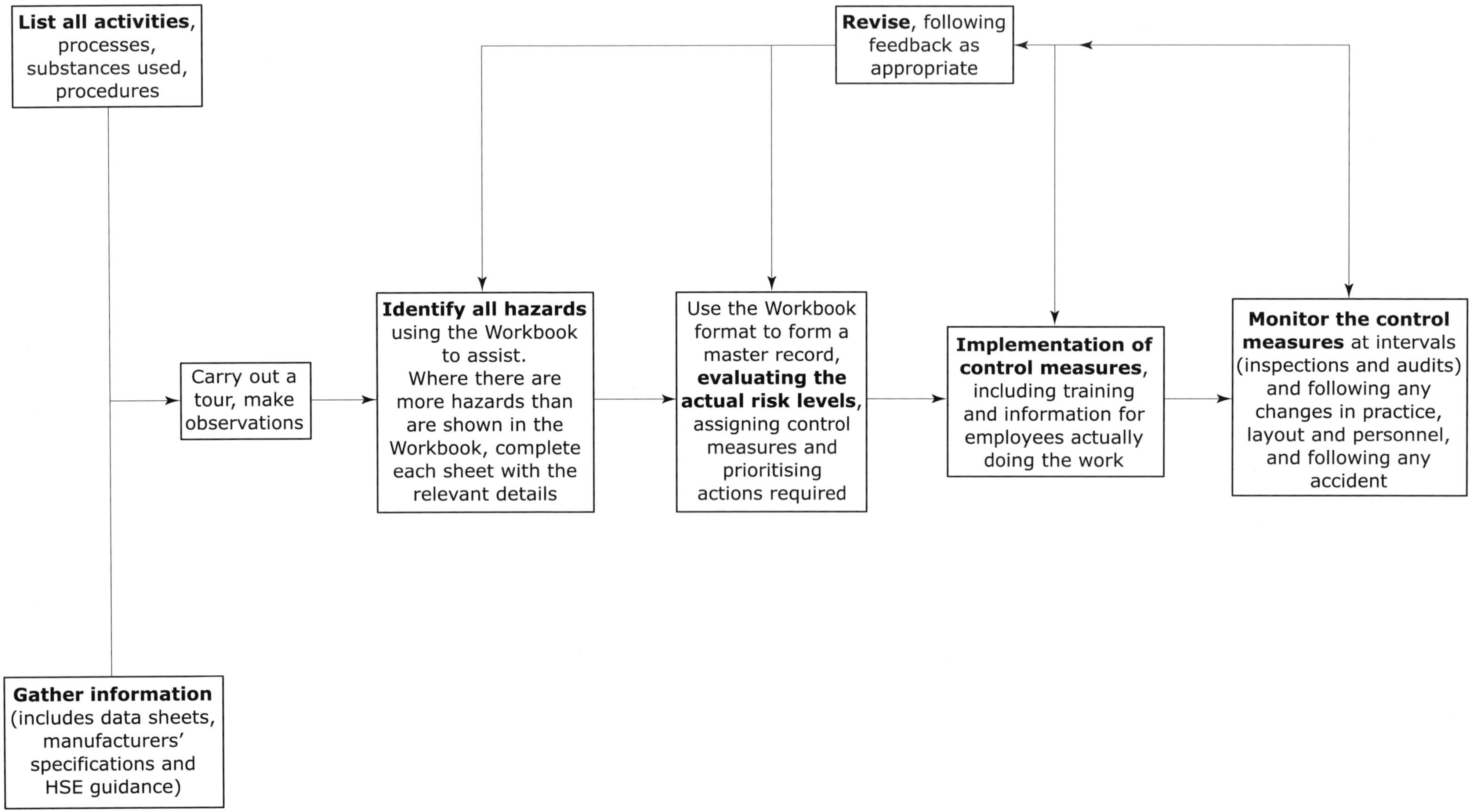

Risk Assessment in the Retail Industry

Risk assessment requirements

Under the *Offices, Shops and Railway Premises Act 1963, s 49*, when a shop owner first begins to employ persons to work in the shop he must inform the appropriate Local Authority (form OSR1). The Act does not apply to premises where only self-employed people work. It is the Local Authority Environmental Health Officer (EHO) who is responsible for enforcing health and safety law in retail premises and not the Health & Safety Executive (HSE).

It is also the duty of the employer to inform employees about health and safety law, either by displaying the statutory poster (*Health & Safety Law – what you should know*) or by giving each employee the approved leaflet.

The poster must state the name and address of the enforcing authority for the premises and the address of the local Employment Medical Advisory Service (EMAS).

The *Management of Health and Safety at Work Regulations 1999 (SI 1999 No 3242)* requires employers to carry out 'suitable and sufficient' risk assessment, to take preventive and protective measures and to record any 'significant findings'.

Industry guidelines

The HSE produces very little health and safety guidance specifically for the retail industry. Four documents that will be of help when considering risk assessment are:

- *Preventing Violence to Retail Staff*, HS(G)165 (ISBN 0 7176 0891 3);
- *Health & Safety in Retail and Wholesale Warehouses*, HS(G)76 (ISBN 0 11 885731 2);
- *Checkouts and musculoskeletal disorders*, INDG 269;
- *Preventing Violence to Retail Staff*, HS(G)133 (ISBN 0 7176 0891 3).

These documents will help to ascertain who is at risk of violence; be it verbal abuse, threats or assault, and procedures that can be implemented to reduce the risk.

Other documents available from HSE books, although not specific to the retail industry, will be useful when risk assessing: 'Working alone in safety; Work-related upper limb disorders – a guide to prevention'. There are also books available from the HSE, which include the actual Regulations for risk assessment and guidance and/or Approved Codes of Practice (ACoPs). Details of those that may prove useful to risk assessors in the industry can be found in the back of this workbook. Other Regulations have been mentioned, which do not specify risk assessment, but are useful management tools when risk assessing. The HSE's website is also useful for gaining information (www.hse.gov.uk).

The TUC's website WorkSMART at www.worksmart.org.uk deals with such issues as:

- violence and assault;
- checkout comfort;
- working in high temperatures; and
- who is responsible for health and safety in shops.

USDAW, the shop workers union, is a valuable source for industry guidelines, and an example of some of the current information available on the USDAW website (www.usdaw.org.uk) includes:

- 'Let's put a stop to bullying at work';

- 'Late-night working – preventing violence to staff';
- 'The Working Time Directive: where do we go from here?;
- .Preventing manual handling injuries';
- 'Checking out health and safety in shops'; and
- 'Keep your cool! Tackling heat stress at work'.

Britain's General Union the GMB produces a leaflet 'Health & Safety in the Retail Sector', which includes common hazards in the retail sector.

The *Disability Discrimination Act 1995* – new requirements to make goods, facilities, services and premises more accessible to disabled people from 2004 for both disabled customers and employees. A number of leaflets are available from the Disability Rights Commission (DRC) including 'Improving access to goods and services for disabled customers – A case study of a clothes shop'. Similar documents are also available for small shops; newsagents and cafes. These can be downloaded from the DRC website:

www.drc-gb.org/drc/InformationAndLegislation/Page.331a.asp

Information relating to the Act can be found on www.disability.gov.uk/dda/index.html. As this is an important subject, with a number of implications for the retail industry it is strongly recommended that the reader familiarises themselves with the contents of the Act.

Industry point of view

As stated above in the section headed INDUSTRY GUIDELINES, it is often Trade Unions that bring information to the notice of employees, and it is noted from the amount of leaflets and other literature concerning the retail sector that there is a general increase in violence to staff. The British Retail Consortium (BRC) often has an industry point of view as shown in their comments to a Consultation Paper on the Safety of Services for Consumers published by the European Commission Directorate General for Health and Consumer Protection. Committees representative of their membership as a whole determine the policies and position of the British Retail Consortium.

Industry anomalies

An HSC (Health and Safety Commission) fact sheet shows that in 1998/99 there were approximately 445,000 retail premises employing around 2.7m people. Employers have a duty to protect these employees and members of the public who visit retail premises.

Environmental Health Officers (EHOs) and Health & Safety Executive Inspectors (HSE) get to hear about most workplace deaths through the *Reporting of Injuries, Diseases and Dangerous Occurrences Regulations 1995 (SI 1995 No 3163) (RIDDOR 1995)*. But a Labour Force Survey (LFS) identified that only 40% of non-fatal injuries to employees in the retail sector are reported.

The HSC Fact Sheet (HELA 2000) details the number of fatalities and injuries to both employees and members of the public reported to Local Authorities under *RIDDOR 1995* by employers and others for a five-year period from 1994/95 to 1998/99.

The statistics cover the retail industry as a whole; during the five-year period, 67,394 Injuries were reported within the industry. Of these, 15 were fatal injuries; 16,875 major injuries; and 50,504 'over-three-day injuries' (where an employee is off work for more than three days due to an injury).

The following Table shows the status of these reported injuries.

Employment status	Severity of injury			
	Fatal	*Major*	*Over-3-day*	*Total*
Employees	7	8,719	50,400	59,126
Self-employed	2	86	104	192
Members of the public	6	8,070	*	8,076
Total	15	16,875	50,504	67,394
* Not reportable				

Note: The seven fatalities to employees represented 12% of all employee fatalities in the Local Authority sector.

Employee fatalities (7)	
Struck by moving vehicle (1 goods vehicle, 2 fork lift truck):	4
Fall from a height:	2
Trapped by a falling or collapsing object:	1

Members of the public fatalities (6)	
Struck by moving object:	2
Struck by moving vehicle:	1
Slip or trip:	1
Fall from a height:	1
Drowning or asphyxiation:	1

Note: Of these fatal injuries to employees, one occurred in a loading bay, one in a car park, one in a stockroom, one in a shopping mall and two in warehouses – we do not have the location of the seventh victim.

Type of accident	Employees *Major*	Employees *Over-3-day*	Public *Non-fatal*
Slip, trip or fall on the same level	3449	10,733	5,082
Handling, lifting or carrying	1217	18,543	
Struck by moving object	1207	8,105	4,000

Note: The location of the largest number of major and over three-day injuries to employees and non-fatal injuries to members of the public was in the 'sales area'.

It is not possible to provide a complete list of areas to risk assess as individual businesses and premises will have different hazards where risks need to be assessed and controlled. However, the following areas listed below indicate the location of accidents to both employees and members of the public, which can be used as a starting point for areas to risk assess.

Sales areas	Warehouse
Stock room	Loading bay
Stairs	Entrance or exit
Canteen/restaurant	Washroom/toilets
Car parks	Delivery off premises
Cold room	Meat room
Offices	Bakery
Preparation room (food)	Kitchen
Area outside building	Staff room
Corridor	Mail order
Shopping mall	Pavement
Skips	Workshops
Window cleaning	Escalators
Lifts	Other places

Terminology

Industry specific

Shop premises – this covers a number of types of premises: (as determined by the *Offices, Shops and Railway Premises Act 1963*):

(a) a shop;

(b) a building or part of a building used principally for retail;

(c) a building occupied by a wholesale dealer, or where goods are kept for wholesale;

(d) a building, which the public use for the delivery of goods for cleaning, repair or treatment. eg dry cleaners or where the public carry out their own treatments such as launderettes;

(e) fuel storage purposes – Premises used for the storage and sale of solid fuel (coal, coke) and related products.

Retail park – a group of buildings set out together and designated for the purpose of selling goods to members of the public. A retail park provides parking, road access, out of town and numerous premises in one location.

Retail premises – examples include the following (not a definitive list):

- shops;
- public houses;
- garages – sale and service of vehicles;
- petrol filling stations;
- department stores;
- warehouses;
- garden centre;
- market stall;
- banks;
- DIY stores;
- restaurants;
- cafés;
- barber or hairdresser;
- wine merchants;
- farm shops;
- pharmacies;
- opticians;
- hotels etc.

Multi-occupancy buildings – a building, which is owned by one person (or one company) and parts of, which are licensed or leased to another, eg a shopping mall. The building has to be of sound construction and properly maintained. A contravention of this will find the owner guilty of an offence.

Franchise – a franchise is a business where the owner has bought the right to use the company name and branding and use the tried and tested methods of the parent company.

Concessions – this is where other retail organisations rent floor space within a department store. There are a number of well-known clothing and cosmetics companies which do this. Other examples include double-glazing companies who have display stands in DIY warehouses or a mobile food/drink unit at a sports event.

Overview of the law regarding risk assessment

Application of legislation in the retail industry

The *Health and Safety at Work etc Act 1974* requires employers to ensure the health, safety and welfare of people at work and others that may come into contact with his undertaking. It also requires the employer to produce a Health and Safety Policy, which will acknowledge the company's commitment to health and safety in a general statement of intent, the organisation for carrying out that intent and arrangements for putting the policy into effect.

The *Management of Health and Safety at Work Regulations 1999 (SI 1999 No 3242)* requires risk assessment, along with other Regulations already mentioned. It is up to Senior Management to ensure that they comply with legislation, if they employ five or more they must appoint a competent person(s) to carry out these duties on their behalf, but they will still be ultimately responsible should anything go wrong. Persons appointed to carry out risk assessment should be trained in the principles and practice.

Examples of good and bad practice

Case study 1: Department store

Bad practice

Soft furnishing department – risk assessment identified that steps leading down to basement were obscured from view by a clutter of stock. Close to the steps was a window with natural light; it had been custom and practice for customers to take goods to this window to view them in natural light. Although a risk assessment had been undertaken, no control measures had been implemented. Consequently, a customer looking to colour match fabrics stepped back from the window and fell backwards down the flight of stairs causing fatal injuries.

Good practice

When carrying out a risk assessment it is imperative if a hazard cannot be removed that control measures are put in place to bring the risk to 'as low as reasonably practicable'.

With the above case study a number of options to manage the risk could be considered:

- Cover the window so that customers do not use it as a viewing platform.
- Remove stock and clutter from vicinity of steps.
- Post notices warning of steps.
- Provide handrails on both sides of steps.
- Provide adequate lighting to area.
- Provide lighting that is the same as natural daylight.

Case study 2: Supermarket

Bad practice

For some shoppers the depth of supermarket trolleys can prove to be a back-breaking experience. Short or tall, retrieving goods from the bottom of the trolley can be painful, especially when leaning over the front or back of the trolley, over-reaching and bending the back can lead to muscular strain and other musculoskeletal disorders.

Employees using those same deep trolleys for taking stock to the shop floor and stacking the shelves will find over a period of time an occasional 'back twinge' may lead to full blown musculoskeletal disorder.

Good practice

Checkout aisles need to be wide enough for customers to be able to stand at the side of the trolley to retrieve goods, thus avoiding leaning from front to back. Trolleys can also be supplied, which are shallower and sometimes wider than the normal deep trolley.

Employees need to have safe systems of work in place to avoid excessive bending, along with manual handling training for safe lifting.

Case study 3: Garden centre

Bad practice

Concrete paving slabs were delivered to the garden centre and lifted one by one onto a trolley by a part-time worker ready to be moved to rear of site. Employee was distracted whilst turning to lay slab onto trolley, dropped slab on his foot and crushed four toes. He was not wearing steel-capped boots, nor had he been provided with manual handling operation training.

Good practice

All such merchandise needs to be delivered on pallets and stored in the position of delivery, not moved by physical effort to the rear of site. PPE to be supplied and worn, eg steel-capped boots and gloves. Manual handling training must be given to all employees, including part-time workers.

Case study 4: Petrol station

Bad practice

A petrol station without a one-way system and a busy forecourt. Two cars enter the petrol station from different directions, both trying to get to the same pump.

Causing a minor head-on collision, both drivers get out of their cars and start shouting at one another. One driver switches on his mobile phone, in doing so, a spark ignites fumes from petrol accidentally spilt by a customer on the other side of pump. At the same time, the other driver, whilst in a state of shock, lights up a cigarette – combustion.

Good practice

The petrol station must have a one-way system to prevent the above scenario; systems in place for immediately cleaning up petrol spills; no smoking signs, sign warning against the use of mobile phones; CCTV security system in place and loudspeaker system allowing cashiers to monitor and control what is happening on the forecourt.

Case study 5: Butchers

Bad practice

A busy butchers' shop, two employees off sick. Butcher alone in shop with a shop full of customers. In a hurry to serve customers he used the bacon slicer without putting guard in place and slices the end off his middle finger.

Good practice

Procedures should be in place so that the guard is always insitu and only lifted off when being cleaned.

Put a sign close to the switch-reminding operator to ensure guard is in place prior to operation.

Slice bacon prior to shop opening or during slack periods. If not already sliced, do not offer sliced meats for sale when short of staff or alone.

A lone worker policy needs to be instigated and PPE (gloves) worn when using equipment.

Case study 6: Corner shop

Bad practice

Owner working in stock room at back of shop early morning prior to staff arriving and shop opening. Using steps to remove merchandise from top shelf, slips on the way down, broken hip, unable to move, had to wait for employee to arrive one hour later.

Good practice

Policy needs to be in place for both lone working and working at a height. Only work at a height when other persons present. Ensure steps suitable for task with non-slip treads.

How to use the workbook and risk assessment methodology

Sample retail company

Modern day health and safety systems focus on risk assessment requiring employers to identify hazards and control any risks evolving from them. But while risk assessment is now a legal requirement, it is not a new concept and can be shown by reviewing two retail premises. One company dates to the 1950/60s and the other in the modern context.

1950s/60s company

The first legislation specifically for shops was the *Shops Act 1950*, which covered such things as early closing, work breaks and Sunday trading etc.

In the late 1950s and early 1960s, prior to the *Offices, Shops and Railway Premises Act 1963*, the *Health & Safety at Work etc Act 1974* and the *Management of Health & Safety at Work Regulations 1992* (now the 1999 Regulations (*SI 1999 No 3242*)), large old department stores where account customers were the wealthy, gentry and aristocracy, risk assessment was not carried out. Manual handling assessments would not have been considered and staff would climb ladders and pick up merchandise without a second thought of possible injury.

The customer was always right and would be given a chair to sit on whilst selecting the goods they required, with junior assistants running around catering to their every demand as well as the department buyer (department manager). The goods would be wrapped and account docket signed. Cash registers were not prominent on the shop floor. Customers paying by cash would have to wait for change while their payment and sales docket were put into a small tube and sent along the mass of pipes to the counting house, where the docket would be receipted and sent back along the pipes with the change. It was the culture of the period.

There are very few such establishments remaining where this sort of service will be found. Commentators may say this is for the better, but there is the loss of the personal touch and stores have moved to the other extreme where very often with the self-service style of purchasing it is difficult to find an assistant willing to give advice.

In some ways, the methods of payment have gone full circle, although the 'store accounts' have been replaced by 'store credit cards', it is very much the same system and alleviates the need for carrying too much cash and so reducing the risk of violence to both staff and customer.

Although risk assessments were not a requirement or formally carried out, 'systems of work' were in place, which helped to reduce the risks, and 'good housekeeping' were key words then as now.

With the general manager doing regular 'walkabouts', nobody would want to incur the wrath of the manager or buyer by having an item out of place. Therefore, faulty displays would have been noticed and reported to maintenance before causing an injury. In the same way, goods and empty boxes etc were not allowed to be left lying around, thus avoiding the tripping hazards. In wet weather porters were tasked with keeping the floor dry and the pavement outside the store free from ice and snow.

Modern-day company

In contrast the following sample company shows the way a department store has evolved with current trends. (NB: This is a hypothetical company and does not reflect a specific organisation.)

Based over five floors, this old, established department store has a wide variety of departments, which include franchise and concessions. Gone are the counters with a number of assistants behind each counter. Open plan is the key with display stands, racks and perhaps only two or three cash desks on each floor – enabling more stock to be displayed.

The layout of the store is as follows:

- basement: china, cookware, glassware, cutlery, gifts, and small electrical appliances;
- ground floor: perfumery, cosmetics, hosiery, stationary, books, leather goods and shoes;
- first floor: ladies' fashions, toilets and café;
- second floor: gents' fashions, sports goods and toys;
- third floor: soft furnishings, pictures, mirrors, lighting and furniture;
- a lift, stairs and escalator run between all floors;
- a despatch department and stock room are at the rear of the ground floor;
- the boiler house is on the third floor, with a water-cooling tower for the air conditioning/heating system on the roof, along with a flagpole;
- access to the roof is gained via a door from the stock room on the third floor; and
- each floor has a behind the scenes stock room.

The store employs approximately 180 staff plus 20 staff employed by the concessions. A health and safety consultant is employed, their role is to monitor and update the safety management system and undertake risk assessment. Accidents and injuries to staff tend to be mainly slips, trips and falls, and manual handling with the odd fall from a height. The falls from a height tend to be stock room employees over-reaching whilst standing on steps. First aiders tend to be used mainly for customers taken ill on the premises. Accidents to customers have, again, consisted of trips and falls.

Although the accident record of the store is generally good, the job of the safety consultant has evolved and more of his time is taken up with health and safety training for employees. It has been decided that every-day risk assessments and daily/weekly monitoring are to be carried out by departmental managers. To this end, managers are being trained to IOSH Managing Safety Standards to enable them to carry out their own risk assessments.

Current risk assessments have identified a need for further manual handling training. New employee induction previously included a video on manual handling operations; this has proven not to be adequate by recent lifting injuries. Therefore, it has been decided that training in lifting and handling of actual goods is to be included in a new training programme.

Risk assessment undertaken during a fire drill identified a lack of basic knowledge as to the whereabouts of fire alarms and assembly areas. It was also identified that fire drills tended to be on a Monday morning when the store was not very busy. Therefore, the drill did not include part-time staff employed only at busy periods. When questioned, these employees were not clear on their duties in an emergency, nor did they know where fire alarm points were located.

During the period of mourning for the Queen Mother, the Union flag was hung at half-mast above the store. After the funeral, members of staff were on the roof to remove the flag. At the same time a police helicopter circling the area identified what was thought to be a group of terrorists on the roof (the flag handlers) and immediately sent in an armed response team. A risk assessment had not been carried out for activities on the roof. Access to the roof is now restricted to selected personnel and local police are notified when this takes place.

Risk assessment workbook

Worked example

Women of child bearing age

TOLLEY'S RISK ASSESSMENT WORKBOOKS – RETAIL INDUSTRY

Topic	Hazard	Risk	People at Risk	Risk Rating	Control Measures	Actioned X	Residual Risk
Women of child bearing age	Physical agents,	Manual handling injuries	Pregnant women and nursing mothers	15	• **Health and safety policy to include procedures for pregnant and nursing mothers.**	☐	4
	Biological and chemical agents	Foetal deformities miscarriage			• **It is unwise for new or expectant mothers to work night shifts. They must provide a signed certificate from a registered medical practitioner or midwife to exclude them from this practice.**	X	
					• **Movements and postures, which can cause physical or mental fatigue, should be avoided.**	☐	
					• **Do not expose pregnant or nursing mothers to chemical or biological agents.**	X	
					• **Alter working conditions or suspend on full pay.**	☐	
						☐	
						☐	
						☐	
						☐	

Owner: AN Other

Assessment date: 01/02/2003
Re-assessment date: 01/08/2003

Worked example (contd)

Women of child bearing age

TOLLEY'S RISK ASSESSMENT WORKBOOKS – RETAIL INDUSTRY

Questions the Risk Assessor needs to ask

No	Question	Yes	No	N/A
1	Do you have written procedures indicating to female workers their obligation to notify you in writing if they become pregnant?	☐	X	☐
2	Are you employing pregnant or nursing mothers on night shifts?	☐	X	☐
3	Are pregnant or nursing mothers involved in manual handling operations?	X	☐	☐
4	Are your pregnant women or nursing mothers likely to be exposed to chemical or biological hazards (eg carbon monoxide fumes, bacteria or viruses)?	☐	X	☐
5	Are you able to alter the working conditions and/or hours worked by pregnant women and nursing mothers?	☐	X	☐
6		☐	☐	☐
7		☐	☐	☐
8		☐	☐	☐
9		☐	☐	☐
10		☐	☐	☐

Owner: AN Other

Assessment date: 01/02/2003
Re-assessment date: 01/03/2003

Worked example (contd)

Women of child bearing age

TOLLEY'S RISK ASSESSMENT WORKBOOKS – RETAIL INDUSTRY

ACTION PLAN

SECTION: Small supermarket

SITE: Queen's Road, Anytown

RISK ASSESSMENT ITEM NO	ACTION REQUIRED	ACTION BY	PRIORITY 1–5	TARGET DATE	COST	COMPLETED (Name/Signature)
1	A review of the Health and Safety Policy must be undertaken to include information for pregnant women and nursing mothers	AN Other	4	22/02/03	Minimal	
3	Pregnant women and nursing mothers must avoid any manual handling operations, ie lifting, pulling, pushing, involving physical effort. Other staff to carry out lifting operations using trolley where possible	AN Other	5	Immediate	Nil	
5	Working conditions and hours cannot be altered; therefore employee must be suspended on full pay and temporary replacement recruited	AN Other	4	22/02/03	Approx £6,000 Wages for temp staff	

KEY	**Action Plan prepared by (name): AN Other**
Priority **1 Monitor** **2 Six months** **3 Two months** **4 Twenty one days** **5 Immediate**	**Signature:** **Date: 01/02/2003**

Useful information

Legislation

- The *Offices, Shops and Railway Premises Act 1963* (most sections have been repealed as a result of new legislation);
- *Health and Safety at Work etc Act 1974*;
- *Management of Health and Safety at Work Regulations 1999 (SI 1999 No 3242)*;
- *Manual Handling Operations Regulations 1992 (SI 1992 No 2793)*;
- *Health and Safety (Display Screen Equipment) Regulations 1992 (SI 1992 No 2792)*
- *Personal Protective Equipment at Work Regulations 1992 (SI 1992 No 2966)*;
- *Control of Substances Hazardous to Health Regulations 2002 (SI 2002 No 2677) (COSHH)*;
- *Noise at Work Regulations 1989(SI 1989 No 1790)*;
- *Control of Asbestos at Work Regulations 2002 (SI 2002 No 2675)*;
- *Workplace (Health, Safety and Welfare) Regulations 1992 (SI 1992 No 3004)*;
- *Provision and Use of Work Equipment Regulations 1998 (SI 1992 No 2306)*;
- *Health and Safety (Miscellaneous Amendments) Regulations 2002 (SI 2002 No 2174)*;
- *Reporting of Injuries, Diseases and Dangerous Occurrences Regulations 1995 (SI 1995 No 3163) (RIDDOR)*;
- *Electricity at Work Regulations 1989 (SI 1989 No 635)*;
- *Health and Safety (Safety Signs and Signals) Regulations 1996 (SI 1996 No 341)*;
- *Health and Safety (First Aid) Regulations 1981 (SI 1981 No 917)*;
- *Lifting Operations and Lifting Equipment Regulations 1998 (SI 1998 No 2307) (LOLER)*;
- *Fire Certificates (Special Premises) Regulations 1976 (SI 1976 No 2003)*;
- *Fire Precautions (Application for a Certificate) Regulations 1989 (SI 1989 No)*;
- *Fire Precautions (Non-certificated Factory, Office, Shop and Railway Premises) (Revocations) Regulations 1989 (SI 1989 No 77)*;
- *Fire Safety and Safety of Places of Sport Act 1987*;
- *Fire Precautions (Workplace) Regulations 1997 (SI 1997 No 1840)* as amended by *SI 1999 No 1877*;
- *Disability Discrimination Act 1995*:
  - *Part I* – Definition of disability;
  - *Part II* – Employment Provisions;
  - *Part II* – Duties of Trade Organisations to their disabled members and applicants;
  - *Part III* – Access to Goods and Services.

Reports

- HELA 2000 HSC 1994/95 TO 1998/99.
- LFS (Labour Force Survey) from Strategy & Analytical support Directorate HSC.

Booklets

Health & Safety Executive

- HS(G)133: 'Preventing Violence to Retail Staff' (ISBN 0 7176 0891 3).
- 'Musculoskeletal disorders in supermarket cashiers' (ISBN 0 7176 0831 X).
- HS(G)136: 'Workplace Transport Safety'.
- HS(G) 76: 'Health & Safety in Retail and Wholesale Warehouses' (ISBN 0 11 885731 2).
- INDG163(rev1): 'Five Steps to Risk Assessment'.

Local Authority Environmental Health Department

- LAC 58/1 'Supermarket checkouts and musculoskeletal disorders'.

Useful addresses

Health & Safety Executive
Rose Court
2 Southwark Bridge Road
London
SE1 9HS

HSE Books
PO Box 1999
Sudbury
Suffolk
CO10 6FS

Institution of Occupational Safety and Health (IOSH)
The Grange
Highfield Drive
Wigston
Leicestershire
LE18 1NN

Chartered Institute of Environmental Health (CIEH)
Chadwick Court
15 Hatfields
London
SE1 8DJ

Part 2

Topic	Hazard	Risk	People at Risk	Risk Rating	Control Measures	Actioned X	Residual Risk
Banking takings	Possible robbery	Major injury			• **Use security company if possible**	☐	
					• **Do not bank at the same time every day**	☐	
					• **Do not go to the bank in a lone worker situation**	☐	
					• **Provide personal safety training for employees responsible for banking**	☐	
					• **Do not make it obvious that money is being carried**	☐	
						☐	
						☐	
						☐	
						☐	
						☐	

Owner:

Assessment date:
Re-assessment date:

Questions the Risk Assessor needs to ask

No	Question	Yes	No	N/A
1	Are you able to employ the services of a security company to collect takings?	☐	☐	☐
2	Can you vary the time that banking takes place?	☐	☐	☐
3	Lone workers are more vulnerable, are you able to send more than one person to the bank?	☐	☐	☐
4	Has personal safety training been made available to employees?	☐	☐	☐
5	Do you allow employees to use a bag that is typical of what they usually carry?	☐	☐	☐
6		☐	☐	☐
7		☐	☐	☐
8		☐	☐	☐
9		☐	☐	☐
10		☐	☐	☐

Owner:

Assessment date:
Re-assessment date:

Topic	Hazard	Risk	People at Risk	Risk Rating	Control Measures	Actioned X	Residual Risk
Boiler room	Fire Explosion Fumes	Personal injury			• **Ensure staff employed to operate boiler are competent** • **Ensure competent contractors regularly service boilers** • **Ensure boiler room fitted with carbon monoxide and fire detectors as appropriate** • **Ensure boiler room access restricted to authorised persons only** • **Ensure boiler room free of obstructions and flammable material**	☐ ☐ ☐ ☐ ☐ ☐ ☐ ☐ ☐ ☐	

Owner:

Assessment date:
Re-assessment date:

Questions the Risk Assessor needs to ask

No	Question	Yes	No	N/A
1	Are staff employed to operate the boiler competent to undertake the task?	☐	☐	☐
2	Is there a system to ensure that a competent contractor regularly services boilers?	☐	☐	☐
3	Is the boiler room fitted with appropriate carbon monoxide and fire detectors?	☐	☐	☐
4	Is access to the boiler room clearly restricted to authorised persons only?	☐	☐	☐
5	Is there a policy to ensure that boiler rooms are free from obstructions, flammable materials or storage?	☐	☐	☐
6		☐	☐	☐
7		☐	☐	☐
8		☐	☐	☐
9		☐	☐	☐
10		☐	☐	☐

Owner:

Assessment date:
Re-assessment date:

Topic	Hazard	Risk	People at Risk	Risk Rating	Control Measures	Actioned X	Residual Risk
Car park	Moving vehicles	Impact injury			• **Place speed restrictions and speed ramps in car park**	☐	
					• **Ensure parking spaces adequately marked**	☐	
					• **Provide specific areas for disabled drivers' vehicles and those with children**	☐	
					• **Provide safe walking within car park for public**	☐	
					• **Control the movement of trolleys in car park**	☐	
						☐	
						☐	
						☐	
						☐	
						☐	

Owner:

Assessment date:
Re-assessment date:

Questions the Risk Assessor needs to ask

No	Question	Yes	No	N/A
1	Is there a speed control system in place in the car park?	☐	☐	☐
2	Are car parking spaces adequately marked with good access and egress?	☐	☐	☐
3	Are there specific parking spaces for disabled drivers' vehicles and for drivers with children?	☐	☐	☐
4	Are protected safe walkways marked out within the car park for the public?	☐	☐	☐
5	Is there a system to control the movement of trolleys left in the car park?	☐	☐	☐
6		☐	☐	☐
7		☐	☐	☐
8		☐	☐	☐
9		☐	☐	☐
10		☐	☐	☐

Owner:

Assessment date:
Re-assessment date:

Topic	Hazard	Risk	People at Risk	Risk Rating	Control Measures	Actioned X	Residual Risk
Catering	Food	Food poisoning Cuts Burns			• **Ensure that food preparation areas are suitable and fit for purpose** • **Ensure that catering staff are competent and have the appropriate certification** • **Ensure that catering staff are provided with PPE (Personal Protective Equipment) and that it is used** • **Ensure that there are hygiene procedures in place and adhered to (to comply with Food Hygiene Regulations)** • **Ensure that where staff/public have access to equipment that it is safe to use**	☐ ☐ ☐ ☐ ☐ ☐ ☐ ☐ ☐ ☐	

Owner:

Assessment date:
Re-assessment date:

Questions the Risk Assessor needs to ask

No	Question	Yes	No	N/A
1	Are all food preparation areas suitable and fit for purpose?	☐	☐	☐
2	Do all food preparation staff have food hygiene certification?	☐	☐	☐
3	Are all staff provided with suitable and sufficient PPE?	☐	☐	☐
4	Are there defined hygiene standards which comply with Food Hygiene Regulations?	☐	☐	☐
5	Where staff or public have access to equipment, eg hot drinks machine, is it safe to use?	☐	☐	☐
6		☐	☐	☐
7		☐	☐	☐
8		☐	☐	☐
9		☐	☐	☐
10		☐	☐	☐

Owner:

Assessment date:
Re-assessment date:

Topic	Hazard	Risk	People at Risk	Risk Rating	Control Measures	Actioned X	Residual Risk
Checkout	Repetitive movements	WRULDs			• **Regular breaks from task**	☐	
					• **Rotate tasks**	☐	
					• **Procedures for reporting problems to management**	☐	
					• **Employees trained to recognise symptoms of WRULDs (work related upper limb disorders)**	☐	
					• **Health surveillance provided**	☐	
						☐	
						☐	
						☐	
						☐	
						☐	

Owner:

Assessment date:
Re-assessment date:

Questions the Risk Assessor needs to ask

No	Question	Yes	No	N/A
1	Do you ensure that employees have regular breaks from repetitive tasks?	☐	☐	☐
2	Are safe systems of work in place to ensure that employees are rotated between different tasks?	☐	☐	☐
3	Are there systems in place where employees can freely report any problems of discomfort to management without feeling that they will be disadvantaged or 'let go' for being considered a 'moaner'?	☐	☐	☐
4	Are all employees trained prior to starting work, not only in the task but also in recognising the risks? (The results of risk assessments must be passed to employees.)	☐	☐	☐
5	Are new employees required to have a health check prior to employment and annual health surveillance to monitor state of health?	☐	☐	☐
6		☐	☐	☐
7		☐	☐	☐
8		☐	☐	☐
9		☐	☐	☐
10		☐	☐	☐

Owner:

Assessment date:
Re-assessment date:

Topic	Hazard	Risk	People at Risk	Risk Rating	Control Measures	Actioned X	Residual Risk
Contract-ors and concess-ions	Workplace	Injury to themselves and others			• **Ensure contractors and concession staff are familiar with emergency procedures and health and safety policy** • **Ensure contractor equipment is suitable and fit for purpose** • **Ensure contractor and concession staff undertake workplace activities in a safe manner** • **Ensure concessions display safety requirements** • **Ensure that work being undertaken by contractors cannot harm staff or public**	☐ ☐ ☐ ☐ ☐ ☐ ☐ ☐ ☐ ☐	

Owner:

Assessment date:
Re-assessment date:

Questions the Risk Assessor needs to ask

No	Question	Yes	No	N/A
1	Is there a system for the familiarisation of contractor and concession staff with emergency procedures and health and safety policy?	☐	☐	☐
2	Is there a system to ensure that contractor's equipment is suitable and fit for purpose whilst in use?	☐	☐	☐
3	Is there a system to monitor that contractors and concession staff undertake their duties in a safe manner?	☐	☐	☐
4	Is there a system to evaluate concession displays - in that they meet safety standards?	☐	☐	☐
5	Are there adequate procedures and systems to ensure that work being undertaken by contractors cannot cause harm to staff or public? (When employing contractors especially for work covered by the Construction Design and Management Regulations you must have a system for ensuring their competence.)	☐	☐	☐
6		☐	☐	☐
7		☐	☐	☐
8		☐	☐	☐
9		☐	☐	☐
10		☐	☐	☐

Owner:

Assessment date:
Re-assessment date:

Topic	Hazard	Risk	People at Risk	Risk Rating	Control Measures	Actioned X	Residual Risk
COSHH	Hazardous substances	Respiratory and other occupational illness			• **Identify all chemicals/substances used in the business. Obtain hazard data sheets from the manufacturer/supplier** • **Suitable and sufficient COSHH assessment to be undertaken on each item** • **Employees to be instructed in the safe use of hazardous substances before use** • **When purchasing substances the safest option for undertaking the activity should be considered** • **Employees to be provided with and trained in the use of personal protective equipment**	☐ ☐ ☐ ☐ ☐ ☐ ☐ ☐ ☐ ☐	

Owner:

Assessment date:
Re-assessment date:

Questions the Risk Assessor needs to ask

No	Question	Yes	No	N/A
1	Have you obtained hazard data sheets for each substance used? (Hazard data sheets should be requested from supplier when purchasing takes place.) (A small sample of hazardous substances, which may be found in retail: general cleaning products, paints, paint stripper, weed killer, solvents.)	☐	☐	☐
2	Are all hazardous substances safe to use with a COSHH assessment completed?	☐	☐	☐
3	Are all staff given instructions on the safe use of chemicals and substances provided by the business?	☐	☐	☐
4	Are all chemicals and substances the safest available for use in the particular workplace?	☐	☐	☐
5	Has an assessment been carried out to identify the type of personal protective equipment required?	☐	☐	☐
6		☐	☐	☐
7		☐	☐	☐
8		☐	☐	☐
9		☐	☐	☐
10		☐	☐	☐

Owner:

Assessment date:
Re-assessment date:

Topic	Hazard	Risk	People at Risk	Risk Rating	Control Measures	Actioned X	Residual Risk
Disabled access	Inadequate access for wheelchairs	Traps, falls	Disabled customers Disabled staff		• **Provide ramp for wheel chair users** • **Ensure door width is adequate (New buildings required minimum door width 800mm for external doors, and 750mm for internal doors. Existing premises 750mm clear opening width.)** • **Train staff to assist disabled customers and colleagues** • **Consider changing the way services are provided to make it easier for disabled customers and employees** • **Position door handles at a suitable height. Or provide a call bell to alert staff when a disabled customer needs assistance**	☐ ☐ ☐ ☐ ☐ ☐ ☐ ☐ ☐ ☐	

Owner:

Assessment date:
Re-assessment date:

Questions the Risk Assessor needs to ask

No	Question	Yes	No	N/A
1	Would it be 'reasonably practicable' to install a ramp where there are steps up to the door?	☐	☐	☐
2	Are your doors (both internal and external) wide enough to accommodate a wheelchair?	☐	☐	☐
3	Has your staff been trained to assist disabled customers and colleagues? (Including those with sensory impairments)	☐	☐	☐
4	Can you provide a home delivery service for disabled customers?	☐	☐	☐
5	Are all door handles at a height manageable by wheelchair users?	☐	☐	☐
6		☐	☐	☐
7		☐	☐	☐
8		☐	☐	☐
9		☐	☐	☐
10		☐	☐	☐

Owner:

Assessment date:
Re-assessment date:

Topic	Hazard	Risk	People at Risk	Risk Rating	Control Measures	Actioned X	Residual Risk
Display of merchan-dise	Falling objects Manual handling	Physical injuries Struck by falling objects			• **Employees trained in display techniques** • **Record kept on personal file** • **Heavy merchandise displayed on lower shelves** • **Ladder training incorporated into manual handling training** • **Assistance given to customers when required**	☐ ☐ ☐ ☐ ☐ ☐ ☐ ☐ ☐ ☐	

Owner:

Assessment date:
Re-assessment date:

Questions the Risk Assessor needs to ask

No	Question	Yes	No	N/A
1	Are employees trained in the techniques required for the safe display of merchandise?	☐	☐	☐
2	Are details of all training kept on employee's personal files?	☐	☐	☐
3	Are staff trained in all aspects of the movement of merchandise using physical effort?	☐	☐	☐
4	Does the training include practical lifting of the type and nature used in the employee's job ie removal of merchandise from high shelving and the use of steps and ladders?	☐	☐	☐
5	Do you ensure that all merchandise is easy for customers to remove from display ie not too high?	☐	☐	☐
6		☐	☐	☐
7		☐	☐	☐
8		☐	☐	☐
9		☐	☐	☐
10		☐	☐	☐

Owner:

Assessment date:
Re-assessment date:

Topic	Hazard	Risk	People at Risk	Risk Rating	Control Measures	Actioned X	Residual Risk
Display stands	Broken/ faulty display stands	Trips, falls and cuts			• **Display stands suitable for purpose**	☐	
					• **Ensure display stands do not have sharp edges or protuberances**	☐	
					• **Ensure display stands are secure and cannot fall over**	☐	
					• **Ensure display stands do not have trip hazard - eg electric power cable**	☐	
					• **Ensure display stands do not block walkway**	☐	
						☐	
						☐	
						☐	
						☐	
						☐	

Owner:

Assessment date:
Re-assessment date:

Questions the Risk Assessor needs to ask

No	Question	Yes	No	N/A
1	Are display stands suitable for the purpose for which they are to be used?	☐	☐	☐
2	Are stands free of sharp edges, broken glass or protuberances that could harm staff or members of the public?	☐	☐	☐
3	Are stands checked to ensure that they are secure and cannot fall over to cause harm?	☐	☐	☐
4	Are stands free of trip hazards such as power cables that may be required for lighting etc?	☐	☐	☐
5	Is there good access between stands? (Eg room for pushchairs and wheelchairs between displays.)	☐	☐	☐
6		☐	☐	☐
7		☐	☐	☐
8		☐	☐	☐
9		☐	☐	☐
10		☐	☐	☐

Owner:

Assessment date:
Re-assessment date:

Topic	Hazard	Risk	People at Risk	Risk Rating	Control Measures	Actioned X	Residual Risk
Doors, manual and automatic	Glass, failure of mechanism	Major personal injury			• **Ensure doors open and close correctly**	☐	
					• **Ensure glass doors have markings, which enable them to be seen**	☐	
					• **Provide automatic doors with an override switch**	☐	
					• **Ensure wooden doors have a glass panel**	☐	
					• **Suitable and working handles to be on all doors**	☐	
						☐	
						☐	
						☐	
						☐	
						☐	

Owner:

Assessment date:
Re-assessment date:

Questions the Risk Assessor needs to ask

No	Question	Yes	No	N/A
1	Do all doors open and close correctly allowing free access in both directions?	☐	☐	☐
2	Have all glass doors got markings at eye level to identify the door to stop people walking into it?	☐	☐	☐
3	Do automatic doors have an override that allows them to operate when no power is available or in an emergency?	☐	☐	☐
4	Do wooden doors have glass panels to allow vision of other side?	☐	☐	☐
5	Do all doors have suitable and working handles on both sides to allow ease of access and use of door?	☐	☐	☐
6		☐	☐	☐
7		☐	☐	☐
8		☐	☐	☐
9		☐	☐	☐
10		☐	☐	☐

Owner:

Assessment date:
Re-assessment date:

Topic	Hazard	Risk	People at Risk	Risk Rating	Control Measures	Actioned X	Residual Risk
Electrical equipment	Electricity	Fatal			• **Ensure all electrical equipment is suitable for purpose**	☐	
	Trailing cables	Electro-cution			• **Ensure all electrical equipment is Portable Appliance Testing (PAT) tested**	☐	
		Burns			• **Ensure electrical equipment has adequate power supply and does not overload system**	☐	
		Trips and falls			• **Systems for the reporting of electrical equipment failure and replacement**	☐	
					• **Ensure staff are trained in the safe use of electrical equipment**	☐	
						☐	
						☐	
						☐	
						☐	
						☐	

Owner:

Assessment date:
Re-assessment date:

Questions the Risk Assessor needs to ask

No	Question	Yes	No	N/A
1	Is all electrical equipment suitable and fit for purpose? (It must comply with CE Standards.)	☐	☐	☐
2	Has all electrical equipment been PAT (Portable Appliance Testing) tested and marked? (At least annually)	☐	☐	☐
3	Are there procedures to identify and monitor power supplies to ensure electrical system is not overloaded?	☐	☐	☐
4	Is there a system in place for the reporting of electrical equipment failure and replacement as well as disposal of broken equipment?	☐	☐	☐
5	Has all staff been trained in the safe use of all electrical equipment?	☐	☐	☐
6		☐	☐	☐
7		☐	☐	☐
8		☐	☐	☐
9		☐	☐	☐
10		☐	☐	☐

Owner:

Assessment date:
Re-assessment date:

Topic	Hazard	Risk	People at Risk	Risk Rating	Control Measures	Actioned X	Residual Risk
Electrical sockets, light fittings and light switches	Electricity	Electro-cution			• **Ensure electrical wiring is in good condition and meets legal requirements**	☐	
					• **Ensure that all electric sockets, light fittings and light switches are in good condition**	☐	
					• **Ensure that electrical sockets are not overloaded**	☐	
					• **Employees to be instructed in the use of electrical appliances and adaptors**	☐	
					• **First aiders trained in treatment of electric shock**	☐	
						☐	
						☐	
						☐	
						☐	
						☐	

Owner:

Assessment date:
Re-assessment date:

Questions the Risk Assessor needs to ask

No	Question	Yes	No	N/A
1	Are all electrical systems compliant with current electrical/building regulations?	☐	☐	☐
2	Have all electric sockets, light fittings and light switches been maintained and tested as being in good condition and fit for purpose?	☐	☐	☐
3	Is there a system to check that electrical sockets are not overloaded with electrical equipment? (Do not plug more than 13 amps into a 13 amp socket).	☐	☐	☐
4	Are employees instructed in the use of electrical appliances, extension cables and adaptors?	☐	☐	☐
5	Are first aiders trained to deal with electric shock victims?	☐	☐	☐
6		☐	☐	☐
7		☐	☐	☐
8		☐	☐	☐
9		☐	☐	☐
10		☐	☐	☐

Owner:

Assessment date:
Re-assessment date:

Topic	Hazard	Risk	People at Risk	Risk Rating	Control Measures	Actioned X	Residual Risk
Emergency action plan	Fire	Major injury			• **Emergency action plan training for employees, with regular fire drills**	☐	
		Burns			• **Walkways to be kept clear and free from obstruction**	☐	
		Asphyxia-tion			• **Ensure fire doors are not locked**	☐	
					• **Fire alarm installed and tested**	☐	
					• **Fire exits marked with statutory signs**	☐	
					• **Fire extinguishers in correct locations and tested annually**	☐	
						☐	
						☐	
						☐	
						☐	

Owner:

Assessment date:
Re-assessment date:

Questions the Risk Assessor needs to ask

No	Question	Yes	No	N/A
1	Does the company have a written emergency action plan?	☐	☐	☐
2	Are all escape routes clear and unobstructed allowing free access?	☐	☐	☐
3	Is there a fire alarm system installed, is it adequate for the premises and do you test it weekly?	☐	☐	☐
4	Are all escape routes and emergency exits identified by statutory signs? (Green safe condition signs)	☐	☐	☐
5	Do you have suitable fire extinguishers?	☐	☐	☐
6		☐	☐	☐
7		☐	☐	☐
8		☐	☐	☐
9		☐	☐	☐
10		☐	☐	☐

Owner:

Assessment date:
Re-assessment date:

Topic	Hazard	Risk	People at Risk	Risk Rating	Control Measures	Actioned X	Residual Risk
Emergency lighting	Power failure	Slips, trips and falls Physical injury			• **Emergency lighting installed and working** • **Emergency lighting tested** • **Escape route signs linked to back-up system** • **Maintenance carried out and recorded** • **Ensure emergency lighting in vulnerable areas, such as stairwells**	☐ ☐ ☐ ☐ ☐ ☐ ☐ ☐ ☐ ☐	

Owner:

Assessment date:
Re-assessment date:

Questions the Risk Assessor needs to ask

No	Question	Yes	No	N/A
1	**Is emergency lighting installed?**	☐	☐	☐
2	**Is the emergency lighting system regularly tested?**	☐	☐	☐
3	**Are all illuminated escape route signs linked to emergency lighting system?**	☐	☐	☐
4	**Does a maintenance programme exist for the emergency lighting?**	☐	☐	☐
5	**Does emergency lighting suitably cover all high-risk areas, such as stairwells and areas with no natural light?**	☐	☐	☐
6		☐	☐	☐
7		☐	☐	☐
8		☐	☐	☐
9		☐	☐	☐
10		☐	☐	☐

Owner:

Assessment date:
Re-assessment date:

Topic	Hazard	Risk	People at Risk	Risk Rating	Control Measures	Actioned X	Residual Risk
Ergonomic assess-ment	Display Screen Equipment, workstation	Work-related upper limb disorders, eye strain			• **Assessment carried out in compliance with DSE Regulations** • **Ensure employees have enough space to perform tasks** • **Employees received information, instruction and training in use of equipment. Record of this and all assessments on personal files** • **Employer to offer eye tests and provide funds for purchase of basic spectacles if required** • **Rectify any non-compliance eg provide adjustable chair**	☐ ☐ ☐ ☐ ☐ ☐ ☐ ☐ ☐ ☐	

Owner:

Assessment date:
Re-assessment date:

Questions the Risk Assessor needs to ask

No	Question	Yes	No	N/A
1	DSE legislation includes 'any alphanumeric or graphic display screen, regardless of the display process involved', which includes till units. Have you carried out a DSE assessment in compliance with the DSE Regulations?	☐	☐	☐
2	Have employees enough space to carry out their tasks?	☐	☐	☐
3	Has all staff received sufficient information, instruction and training to enable safe performance of duties?	☐	☐	☐
4	Have you informed staff of their entitlement to free eye tests and basic lenses and frames where required?	☐	☐	☐
5	Are there procedures for employees to report any equipment/workstation faults in between assessments?	☐	☐	☐
6		☐	☐	☐
7		☐	☐	☐
8		☐	☐	☐
9		☐	☐	☐
10		☐	☐	☐

Owner:

Assessment date:
Re-assessment date:

Topic	Hazard	Risk	People at Risk	Risk Rating	Control Measures	Actioned X	Residual Risk
Escalators	Escalator	Trips			• **Ensure escalators are in good working order and operating at the correct speed**	☐	
		Falls			• **Ensure emergency stop fitted and in working order. Warning signs should be in place to help prevent trapping injuries eg pushchairs not allowed, adults to control children etc**	☐	
		Trap			• **Ensure that step treads are in good condition and not damaged**	☐	
					• **Ensure that there are no protuberances to cause trapping, snagging or cutting**	☐	
					• **Ensure that there is a system to restrict access when not in use**	☐	
						☐	
						☐	
						☐	
						☐	
						☐	

Owner:

Assessment date:
Re-assessment date:

Questions the Risk Assessor needs to ask

No	Question	Yes	No	N/A
1	Is there a procedure to ensure that an escalator is in good working order, operating at the correct speed and has maintenance records?	☐	☐	☐
2	Are emergency stops fitted, visible and in good condition?	☐	☐	☐
3	Is there an inspection procedure to ensure that the step treads are in good condition and not damaged?	☐	☐	☐
4	Is there a monitoring system to ensure that escalators are free from trapping, snagging or cutting hazards?	☐	☐	☐
5	When maintenance is being carried out or there is a breakdown, is there a system in place to restrict access? Eg blocking off at top and bottom and warning signs.	☐	☐	☐
6		☐	☐	☐
7		☐	☐	☐
8		☐	☐	☐
9		☐	☐	☐
10		☐	☐	☐

Owner:

Assessment date:
Re-assessment date:

Topic	Hazard	Risk	People at Risk	Risk Rating	Control Measures	Actioned X	Residual Risk
First aid	Injury to staff, public	Injury and illness			• **A suitable first aid kit (equipment) and accident book is a legal requirement** • **There must be a sufficient number of staff trained in first aid or 'appointed persons' nominated** • **There is a requirement for a system to check the first aid kit contents** • **There needs to be a defined policy for staff administering first aid, with suitable PPE (Personal Protective Equipment) provided** • **First aid kits should be readily accessible to staff**	☐ ☐ ☐ ☐ ☐ ☐ ☐ ☐ ☐ ☐	

Owner:

Assessment date:
Re-assessment date:

Questions the Risk Assessor needs to ask

No	Question	Yes	No	N/A
1	Is there a suitable first aid kit available to meet legal requirements with an accident book used to maintain records?	☐	☐	☐
2	Are there a sufficient number of competent first aid trained staff?	☐	☐	☐
3	Is there a member of staff responsible for re-stocking the first aid kit?	☐	☐	☐
4	Is there a policy for staff when administering first aid to include PPE, which must be provided?	☐	☐	☐
5	Is there a statutory sign indicating the name and whereabouts of the first aiders and the first aid box?	☐	☐	☐
6		☐	☐	☐
7		☐	☐	☐
8		☐	☐	☐
9		☐	☐	☐
10		☐	☐	☐

Owner:

Assessment date:
Re-assessment date:

Topic	Hazard	Risk	People at Risk	Risk Rating	Control Measures	Actioned X	Residual Risk
Floor cleaning	Slips, trips and falls COSHH	Personal injury Occupa-tional health			• **COSHH assess all cleaning materials** • **Ensure cleaners are provided with personal protective equipment and use it** • **Ensure cleaning equipment is safe and fit for purpose** • **Procedure and equipment to deal with spills or broken objects** • **Warning signs provided and used when floor-cleaning being carried out**	☐ ☐ ☐ ☐ ☐ ☐ ☐ ☐ ☐ ☐	

Owner:

Assessment date:
Re-assessment date:

Questions the Risk Assessor needs to ask

No	Question	Yes	No	N/A
1	Has a COSHH assessment been carried out for all floor cleaning materials?	☐	☐	☐
2	Have all cleaning staff been provided with suitable and sufficient personal protective equipment?	☐	☐	☐
3	Is all cleaning equipment (electrical) fit for purpose and has it been tested?	☐	☐	☐
4	Is there a procedure in place to clean up spills or broken objects such as glass?	☐	☐	☐
5	Are warning signs provided and used when floor cleaning is being carried out?	☐	☐	☐
6		☐	☐	☐
7		☐	☐	☐
8		☐	☐	☐
9		☐	☐	☐
10		☐	☐	☐

Owner:

Assessment date:
Re-assessment date:

Topic	Hazard	Risk	People at Risk	Risk Rating	Control Measures	Actioned X	Residual Risk
Floors	Slips, trips and falls	Personal injury			• **Ensure carpets and other floor coverings are in good condition and secure**	☐	
					• **Ensure tiled floors are in good condition and secure**	☐	
					• **Ensure that there are no slip hazards**	☐	
					• **Ensure that there are no trip hazards (ie trailing leads)**	☐	
					• **Ensure that tile/wood floors are non-slip and kept free of water**	☐	
						☐	
						☐	
						☐	
						☐	
						☐	

Owner:

Assessment date:
Re-assessment date:

Questions the Risk Assessor needs to ask

No	Question	Yes	No	N/A
1	Are you able to employ the services of a security company to collect takings?	☐	☐	☐
2	Do you have procedures in place for notifying damaged floorings and undertaking repairs or replacement?	☐	☐	☐
3	Are procedures in place for removing all spillages immediately?	☐	☐	☐
4	Are floors clear of other trip hazards such as cables and goods?	☐	☐	☐
5	Tile and wood floors are susceptible to slips when wet. Are there procedures to provide for a non-slip surface and keep dry?	☐	☐	☐
6		☐	☐	☐
7		☐	☐	☐
8		☐	☐	☐
9		☐	☐	☐
10		☐	☐	☐

Owner:

Assessment date:
Re-assessment date:

Topic	Hazard	Risk	People at Risk	Risk Rating	Control Measures	Actioned X	Residual Risk
Ladders Extending ladders Step ladders	Falls from a height	Major injury			• **Industrial quality stepladders should be used, a ladder register should be kept** • **Ladders should be portable and suitable for the purpose to be used** • **Ladders should be fitted with anti-tilt device or security system** • **There needs to be a system to ensure floor/ground is suitable for the safe use of ladder/step ladder** • **The appropriate number of staff needs to be used to ensure safe use of ladder/step ladder**	☐ ☐ ☐ ☐ ☐ ☐ ☐ ☐ ☐ ☐	

Owner:

Assessment date:
Re-assessment date:

Questions the Risk Assessor needs to ask

No	Question	Yes	No	N/A
1	Is there a system to ensure that only industrial quality ladders/step ladders are used?	☐	☐	☐
2	Are all ladders used suitable and fit for the purpose intended and is there a maintenance record kept?	☐	☐	☐
3	Are all extending ladders fitted with an anti-tilting device and/or a securing system?	☐	☐	☐
4	Is there a procedure/policy to ensure that the floor and/or ground are suitable for the safe use of ladder or steps?	☐	☐	☐
5	Is there a policy to determine the number of staff to be used to ensure safe use of extending ladders or stepladder?	☐	☐	☐
6		☐	☐	☐
7		☐	☐	☐
8		☐	☐	☐
9		☐	☐	☐
10		☐	☐	☐

Owner:

Assessment date:
Re-assessment date:

Topic	Hazard	Risk	People at Risk	Risk Rating	Control Measures	Actioned X	Residual Risk
Lifts	Lift car, lift shaft	Trap			• **Ensure that lifts are fit for purpose and have in-date test**	☐	
		Fall			• **Ensure that there are emergency procedures in the event of a failure, which should include the call out of emergency services and/or lift engineer - staff should not try to rescue trapped person(s)**	☐	
		Fire			• **Ensure adequate notices for number of people to be carried**	☐	
					• **Ensure emergency alarm/phone works**	☐	
					• **Ensure system to stop use of lift in event of fire eg display warning sign 'Do not use lift in an emergency'**	☐	
						☐	
						☐	
						☐	
						☐	
						☐	

Owner:

Assessment date:
Re-assessment date:

Questions the Risk Assessor needs to ask

No	Question	Yes	No	N/A
1	Do you have a maintenance schedule for your lift?	☐	☐	☐
2	Are there emergency procedures in place to deal with a lift failure and trapped persons?	☐	☐	☐
3	Are there notices to determine the safe passenger load that can be seen and understood by passengers?	☐	☐	☐
4	Is there an emergency phone/alarm fitted to the lift?	☐	☐	☐
5	Is there a procedure to stop a lift being used in the event of a fire and to evacuate persons if necessary?	☐	☐	☐
6		☐	☐	☐
7		☐	☐	☐
8		☐	☐	☐
9		☐	☐	☐
10		☐	☐	☐

Owner:

Assessment date:
Re-assessment date:

Topic	Hazard	Risk	People at Risk	Risk Rating	Control Measures	Actioned X	Residual Risk
Lift truck	Moving vehicles	Collision			• **Ensure staff are competent to use lift trucks**	☐	
		Falling loads			• **Ensure lift truck is in good working condition and fit for the purpose**	☐	
		Fire			• **Ensure adequate protection for lift truck drivers**	☐	
		CO poisoning			• **Ensure adequate protection to buildings and pedestrians from vehicle movement eg safe working lanes and areas for truck movement**	☐	
					• **Procedures in place for battery charging and the use of combustion engines in confined spaces**	☐	
						☐	
						☐	
						☐	
						☐	
						☐	

Owner:

Assessment date:
Re-assessment date:

Questions the Risk Assessor needs to ask

No	Question	Yes	No	N/A
1	Is there a system in place to ensure that lift truck drivers are competent? (Training certificate should be kept on personal files.)	☐	☐	☐
2	Is there a system to ensure that lift trucks are in good condition, fit for purpose and have a maintenance record?	☐	☐	☐
3	Is the lift truck fitted with suitable protection to ensure that the driver is safe, ie a harness/seat belt?	☐	☐	☐
4	Are there working lanes and areas for the lift truck to operate which are marked with warning signs?	☐	☐	☐
5	Do you have a no smoking policy in place in the area of battery charging?	☐	☐	☐
6		☐	☐	☐
7		☐	☐	☐
8		☐	☐	☐
9		☐	☐	☐
10		☐	☐	☐

Owner:

Assessment date:
Re-assessment date:

Topic	Hazard	Risk	People at Risk	Risk Rating	Control Measures	Actioned X	Residual Risk
Loading bays	Moving vehicles	Contact with property, pedestrians			• **Hatched markings on ground** • **Barriers around loading bay** • **Refuge provided** • **Banksman used for reversing vehicles** • **Parking restrictions in loading bay**	☐ ☐ ☐ ☐ ☐ ☐ ☐ ☐ ☐ ☐	

Owner:

Assessment date:
Re-assessment date:

Questions the Risk Assessor needs to ask

No	Question	Yes	No	N/A
1	Do you have pedestrian and vehicle routes clearly marked? Is there a procedure for re-marking hatched lines when faded? Do you have statutory signs warning pedestrians of moving vehicles?	☐	☐	☐
2	Are barriers erected around loading bays to prevent pedestrian access?	☐	☐	☐
3	Alternatively can you provide a refuge which can be used to avoid being struck or crushed by moving vehicles?	☐	☐	☐
4	Are banksmen used to aid the drivers of reversing vehicles?	☐	☐	☐
5	Are warning notices displayed to prevent illegal parking in loading bays?	☐	☐	☐
6		☐	☐	☐
7		☐	☐	☐
8		☐	☐	☐
9		☐	☐	☐
10		☐	☐	☐

Owner:

Assessment date:
Re-assessment date:

Topic	Hazard	Risk	People at Risk	Risk Rating	Control Measures	Actioned X	Residual Risk
Lone worker	Lone working	Fatal injury Accident Abductions			• **Ensure staff working alone are competent and have adequate and suitable equipment** • **Ensure staff have a system for summoning assistance in an emergency** • **Ensure systems to reduce risk of harm to lone workers** • **Ensure that lone staff can safely undertake work** • **Ensure a system to monitor the location of travelling lone workers eg delivery staff**	☐ ☐ ☐ ☐ ☐ ☐ ☐ ☐ ☐ ☐	

Owner:

Assessment date:
Re-assessment date:

Questions the Risk Assessor needs to ask

No	Question	Yes	No	N/A
1	**Are all staff that work alone competent and do they have adequate and suitable workplace equipment?**	☐	☐	☐
2	**Do all staff have a system for summoning assistance in the event of an emergency?**	☐	☐	☐
3	**Are safe systems of work in place for lone working?**	☐	☐	☐
4	**Is there a system to check that lone working staff can safely undertake the work?**	☐	☐	☐
5	**Is there a system that monitors the location of lone workers who are travelling as part of their workplace activities?**	☐	☐	☐
6		☐	☐	☐
7		☐	☐	☐
8		☐	☐	☐
9		☐	☐	☐
10		☐	☐	☐

Owner:

Assessment date:
Re-assessment date:

Topic	Hazard	Risk	People at Risk	Risk Rating	Control Measures	Actioned X	Residual Risk
Portable workplace equipment Equipment not normally present	Work equipment	Trips			• **Portable equipment used in workplace must be subjected to control procedures**	☐	
		Falls			• **Portable equipment must be suitable, fit for purpose and tested (if appropriate)**	☐	
		Cuts			• **The use of portable equipment may require warning signs**	☐	
		Burns			• **Portable fires, electric or gas, fans need to be fit for purpose and located in a safe position**	☐	
		Electro-cutions			• **Does portable electric equipment have an adequate and suitable power supply?**	☐	
						☐	
						☐	
						☐	
						☐	
						☐	

Owner:

Assessment date:
Re-assessment date:

Questions the Risk Assessor needs to ask

No	Question	Yes	No	N/A
1	Is there a system to ensure that portable equipment is suitable and fit for purpose? Is all portable electrical equipment PAT tested?	☐	☐	☐
2	Is there a requirement for warning signs to be used when portable equipment is used?	☐	☐	☐
3	If portable fires, either gas or electric, or fans are used are they fit for purpose and located in a safe place with no risk of harm to persons?	☐	☐	☐
4	When portable electric equipment is used are extension leads kept to a minimum?	☐	☐	☐
5		☐	☐	☐
6		☐	☐	☐
7		☐	☐	☐
8		☐	☐	☐
9		☐	☐	☐
10		☐	☐	☐

Owner:

Assessment date:
Re-assessment date:

Topic	Hazard	Risk	People at Risk	Risk Rating	Control Measures	Actioned X	Residual Risk
Premises	Damaged asbestos	Occupa-tional health Inhalation Skin contact			• **Survey to be carried out of all known asbestos by registered contractor. Asbestos sealed, tagged and logged** • **Procedures for the discovery of materials or dust, which may contain asbestos** • **Asbestos may only be worked upon by competent person and removed by authorised, registered contractors** • **Employees informed of location of asbestos**	☐ ☐ ☐ ☐ ☐ ☐ ☐ ☐ ☐ ☐	

Owner:

Assessment date:
Re-assessment date:

Questions the Risk Assessor needs to ask

No	Question	Yes	No	N/A
1	Have you had a survey completed by an authorised registered contractor?	☐	☐	☐
2	Do you have a site register for the location, type and condition of asbestos?	☐	☐	☐
3	Are employees informed of the procedures laid down in the event that they think there is asbestos present?	☐	☐	☐
4	If asbestos is discovered, do you have details of an authorised, registered contractor?	☐	☐	☐
5	When identified, are employees informed of the location of asbestos?	☐	☐	☐
6		☐	☐	☐
7		☐	☐	☐
8		☐	☐	☐
9		☐	☐	☐
10		☐	☐	☐

Owner:

Assessment date:
Re-assessment date:

Topic	Hazard	Risk	People at Risk	Risk Rating	Control Measures	Actioned X	Residual Risk
Sale of knives	Sale of knives to young people	Personal injury Injury to others			• **Procedures in place for refusing the sale of knives to children (under 16)**	☐	
					• **Knife refusal register at each checkout**	☐	
					• **Knives with long blades displayed in locked cabinet**	☐	
					• **All checkout staff aware of procedures**	☐	
					• **Notices displayed warning of procedures for sale of knives**	☐	
						☐	
						☐	
						☐	
						☐	
						☐	

Owner:

Assessment date:
Re-assessment date:

Questions the Risk Assessor needs to ask

No	Question	Yes	No	N/A
1	Legislation does not permit the sale of knives to children - is there a policy in place to stop this happening?	☐	☐	☐
2	Is there a completed knife refusal register at each checkout?	☐	☐	☐
3	Do you have lockable cabinets for the display of knives?	☐	☐	☐
4	Are all checkout staff given training in all procedures at induction, including the use of the knife refusal procedures and register?	☐	☐	☐
5	Do you have warning signs displayed to deter children from trying to purchase knives?	☐	☐	☐
6		☐	☐	☐
7		☐	☐	☐
8		☐	☐	☐
9		☐	☐	☐
10		☐	☐	☐

Owner:

Assessment date:
Re-assessment date:

Topic	Hazard	Risk	People at Risk	Risk Rating	Control Measures	Actioned X	Residual Risk
Sale of solvents (glue, lighter fuel etc)	Sale of solvents	Solvent abuse Asphyxia-tion			• **Solvent refusal register at each checkout** • **Procedures in place for refusing the sale of solvents** • **All checkout/sales staff aware of procedures** • **Solvents locked in display cabinet** • **Notices displayed to warn would-be solvent abusers against trying to purchase products**	☐ ☐ ☐ ☐ ☐ ☐ ☐ ☐ ☐ ☐	

Owner:

Assessment date:
Re-assessment date:

Questions the Risk Assessor needs to ask

No	Question	Yes	No	N/A
1	Do you have procedures in place to deter the purchase of such products in order to prevent solvent abuse by young people?	☐	☐	☐
2	Is there a solvent refusal register at each checkout, is it completed and regularly monitored?	☐	☐	☐
3	Are all solvents displayed in a locked cabinet?	☐	☐	☐
4	Are all checkout/sales staff familiar with procedures?	☐	☐	☐
5	Are notices displayed warning young people against trying to purchase solvents?	☐	☐	☐
6		☐	☐	☐
7		☐	☐	☐
8		☐	☐	☐
9		☐	☐	☐
10		☐	☐	☐

Owner:

Assessment date:
Re-assessment date:

Topic	Hazard	Risk	People at Risk	Risk Rating	Control Measures	Actioned X	Residual Risk
Security	Theft	Personal injury			• **Policy and systems for dealing with theft**	☐	
					• **Staff trained to deal with potential acts of theft**	☐	
					• **System for staff to notify security staff of potential theft**	☐	
					• **System to have goods security coded or marked.**	☐	
					• **Policy to minimise risk or theft - minimum cash at checkouts**	☐	
						☐	
						☐	
						☐	
						☐	
						☐	

Owner:

Assessment date:
Re-assessment date:

Questions the Risk Assessor needs to ask

No	Question	Yes	No	N/A
1	Is there a company policy for dealing with theft by those not employed?	☐	☐	☐
2	Are staff trained to deal with potential acts of theft if no security staff available or employed? (Exposure to threat)	☐	☐	☐
3	If security staff are employed is there a system for staff to alert security to acts or potential acts of theft? (Non-exposure to threat)	☐	☐	☐
4	Are all goods security coded or marked to reduce the risk of theft and potential threat to staff?	☐	☐	☐
5	Is there a policy of minimising the actions of theft by having limited cash at checkouts?	☐	☐	☐
6		☐	☐	☐
7		☐	☐	☐
8		☐	☐	☐
9		☐	☐	☐
10		☐	☐	☐

Owner:

Assessment date:
Re-assessment date:

Topic	Hazard	Risk	People at Risk	Risk Rating	Control Measures	Actioned X	Residual Risk
Security	Violence	Personal injury			• **Policy and safe systems of work for preventing violence against staff** • **Staff trained to deal with violent/aggressive customers/clients** • **Procedures to reduce potential theft ie till emptied regularly throughout the day to avoid a build-up of cash** • **CCTV system installed and working** • **Trained staff available to deal with security matters**	☐ ☐ ☐ ☐ ☐ ☐ ☐ ☐ ☐ ☐	

Owner:

Assessment date:
Re-assessment date:

Questions the Risk Assessor needs to ask

No	Question	Yes	No	N/A
1	Is there a policy in place for dealing with violence and aggressive behaviour towards staff?	☐	☐	☐
2	Is cash removed from the till to a secure place throughout the day?	☐	☐	☐
3	Are there procedures to reduce the potential of threats and violence against staff? (Eg restricted access to money or goods.)	☐	☐	☐
4	Is there a CCTV system installed?	☐	☐	☐
5	In areas of potential risk (high value goods, cash) are trained security staff available to take control of confrontational situations?	☐	☐	☐
6		☐	☐	☐
7		☐	☐	☐
8		☐	☐	☐
9		☐	☐	☐
10		☐	☐	☐

Owner:

Assessment date:
Re-assessment date:

Topic	Hazard	Risk	People at Risk	Risk Rating	Control Measures	Actioned X	Residual Risk
Service cupboards	Gas	Fire			• **Clear access to cupboard maintained**	☐	
	Electricity	Electric shock			• **Cupboard kept locked when not in use**	☐	
	Water	Flooding			• **Safety notices displayed**	☐	
					• **Access by designated staff only**	☐	
					• **Use for general storage prohibited**	☐	
						☐	
						☐	
						☐	
						☐	
						☐	

Owner:

Assessment date:
Re-assessment date:

Questions the Risk Assessor needs to ask

No	Question	Yes	No	N/A
1	**Is clear access to the services cupboard maintained?**	☐	☐	☐
2	**Is the services cupboard kept locked? (The key must be easily accessible in the event of an emergency.)**	☐	☐	☐
3	**Is a safety notice displayed to designate gas and electrical mains?**	☐	☐	☐
4	**Is access limited to designated personnel?**	☐	☐	☐
5	**Have you ascertained that service cupboards are not used for general storage?**	☐	☐	☐
6		☐	☐	☐
7		☐	☐	☐
8		☐	☐	☐
9		☐	☐	☐
10		☐	☐	☐

Owner:

Assessment date:
Re-assessment date:

Topic	Hazard	Risk	People at Risk	Risk Rating	Control Measures	Actioned X	Residual Risk
Shop entrance/ exit	Wet floor, ice or snow, fallen leaves	Slips, trips and falls			• **Non-slip matting at entrance/exit**	☐	
					• **Ice and snow cleared from entrances**	☐	
					• **Fallen leaves to be removed from footpaths**	☐	
					• **Procedures in place for mopping wet floor**	☐	
					• **Handrail in place on ramps/steps leading to entrance**	☐	
						☐	
						☐	
						☐	
						☐	
						☐	

Owner:

Assessment date:
Re-assessment date:

Questions the Risk Assessor needs to ask

No	Question	Yes	No	N/A
1	Do you have non-slip matting in all doorways?	☐	☐	☐
2	Are there procedures/safe systems of work and personnel identified to clear ice and snow from all entrances/ exits?	☐	☐	☐
3	Are there procedures/safe systems of work and personnel identified to remove fallen leaves from all entrances/ exits, footpaths, steps and ramps?	☐	☐	☐
4	Are there procedures/safe systems of work and personnel identified to mop wet floors?	☐	☐	☐
5	If you have ramps or steps leading to the entrances/exits is there a handrail in place?	☐	☐	☐
6		☐	☐	☐
7		☐	☐	☐
8		☐	☐	☐
9		☐	☐	☐
10		☐	☐	☐

Owner:

Assessment date:
Re-assessment date:

Topic	Hazard	Risk	People at Risk	Risk Rating	Control Measures	Actioned X	Residual Risk
Signs	Lack of warning signs	Personal injury Damage to property			• **Ensure that appropriate statutory signs are in place**	☐	
					• **Ensure that signs conform to the Regulations**	☐	
					• **Ensure where appropriate that illuminated signs are used**	☐	
					• **Ensure that staff understands the signs**	☐	
					• **Ensure signs are used to maximum effect**	☐	
						☐	
						☐	
						☐	
						☐	
						☐	

Owner:

Assessment date:
Re-assessment date:

Questions the Risk Assessor needs to ask

No	Question	Yes	No	N/A
1	Are the appropriate statutory signs in place? (Eg warning, prohibition, mandatory and safe condition.)	☐	☐	☐
2	Do the signs comply with the requirements of the regulations? (Health and Safety (Safety Signs and Signals) Regulations 1996 (SI 1996 No 341))	☐	☐	☐
3	Are illuminated signs fitted where appropriate?	☐	☐	☐
4	Do staff understand the signs and what is required?	☐	☐	☐
5	Are only essential signs used to obtain maximum effect?	☐	☐	☐
6		☐	☐	☐
7		☐	☐	☐
8		☐	☐	☐
9		☐	☐	☐
10		☐	☐	☐

Owner:

Assessment date:
Re-assessment date:

Topic	Hazard	Risk	People at Risk	Risk Rating	Control Measures	Actioned X	Residual Risk
Smoking/ non-smoking areas	Passive smoking	Ill health			• **Designated smoking area provided**	☐	
					• **Adequate ventilation supplied**	☐	
					• **Staff informed of non-smoking areas**	☐	
					• **Heat detectors provided in smoking area**	☐	
					• **Adequate ashtrays provided and emptied regularly**	☐	
					• **No smoking signs displayed in appropriate areas**	☐	
						☐	
						☐	
						☐	
						☐	

Owner:

Assessment date:
Re-assessment date:

Questions the Risk Assessor needs to ask

No	Question	Yes	No	N/A
1	Are staff informed of non-smoking areas? (All employees have a right not to enter the smoking area.)	☐	☐	☐
2	Have you ensured that your ventilation system does not transfer smoke from smoking area to non-smoking areas?	☐	☐	☐
3	Do you provide fire extinguishers in the smoking area?	☐	☐	☐
4	Are ashtrays emptied regularly?	☐	☐	☐
5	Are no smoking signs displayed in all non-smoking areas?	☐	☐	☐
6		☐	☐	☐
7		☐	☐	☐
8		☐	☐	☐
9		☐	☐	☐
10		☐	☐	☐

Owner:

Assessment date:
Re-assessment date:

Topic	Hazard	Risk	People at Risk	Risk Rating	Control Measures	Actioned X	Residual Risk
Staff health and safety training	Lack of training	Personal injury to self or others			• **Ensure that all staff receive health and safety induction** • **Staff to be made aware of risk assessment outcomes** • **Ensure all staff are trained in emergency procedures** • **Staff to receive specific health & safety training as required** • **Ensure part-time staff and concession staff have health and safety induction**	☐ ☐ ☐ ☐ ☐ ☐ ☐ ☐ ☐ ☐	

Owner:

Assessment date:
Re-assessment date:

Questions the Risk Assessor needs to ask

No	Question	Yes	No	N/A
1	Is there a policy to ensure that all staff attend a health and safety induction?	☐	☐	☐
2	Is there a policy and system for staff to be made aware of the risk assessment and procedures in place?	☐	☐	☐
3	Are there procedures to ensure that all staff are trained in emergency actions and exercises?	☐	☐	☐
4	Is there a system to identify specific training requirements eg manual handling, COSHH, DSE etc?	☐	☐	☐
5	Is there a policy to ensure that part-time and concession staff undertake a health and safety induction?	☐	☐	☐
6		☐	☐	☐
7		☐	☐	☐
8		☐	☐	☐
9		☐	☐	☐
10		☐	☐	☐

Owner:

Assessment date:
Re-assessment date:

Topic	Hazard	Risk	People at Risk	Risk Rating	Control Measures	Actioned X	Residual Risk
Stairs	Slips, trips and falls	Major injury			• **Ensure all stairs within the business are suitable and in good condition**	☐	
					• **Ensure carpeting, tiles, other covering or wood are secure and in good condition**	☐	
					• **Ensure banisters and safety rails are fitted and in good, secure condition**	☐	
					• **Ensure stairs are suitable for public access eg too steep, narrow or open**	☐	
					• **Ensure stairs are suitable and safe for children - warning signs if appropriate**	☐	
						☐	
						☐	
						☐	
						☐	
						☐	

Owner:

Assessment date:
Re-assessment date:

Questions the Risk Assessor needs to ask

No	Question	Yes	No	N/A
1	Are all stairs suitable for use and in good condition (eg ensure metal edgings are not sharp)?	☐	☐	☐
2	Is the floor covering on stairs secure and in good condition?	☐	☐	☐
3	Do the stairs have secure banisters and handrails?	☐	☐	☐
4	Are stairs suitable for public access (in that they are not too steep, too narrow, well lit)?	☐	☐	☐
5	Are stairs to be used by the public suitable for children eg there are no gaps to fall through?	☐	☐	☐
6		☐	☐	☐
7		☐	☐	☐
8		☐	☐	☐
9		☐	☐	☐
10		☐	☐	☐

Owner:

Assessment date:
Re-assessment date:

Topic	Hazard	Risk	People at Risk	Risk Rating	Control Measures	Actioned X	Residual Risk
Steps in old buildings (uneven, too high)	Trips and falls	Personal injury			• **Make the step visible to those who have access (different colour)**	☐	
					• **Ensure edges of steps have a safe and secure edging**	☐	
					• **If appropriate provide warning signs**	☐	
					• **Identify if ramps would be more appropriate than steps**	☐	
					• **Ensure handrails are in place on all stairways**	☐	
						☐	
						☐	
						☐	
						☐	
						☐	

Owner:

Assessment date:
Re-assessment date:

Questions the Risk Assessor needs to ask

No	Question	Yes	No	N/A
1	Has consideration been given to make steps visible to all of those who have access?	☐	☐	☐
2	Do all steps have a suitable, safe and secure edging fitted?	☐	☐	☐
3	Is it appropriate and possible to install warning signs on steps that pose a high risk?	☐	☐	☐
4	Has consideration been given to installing ramps to eliminate steps, particularly if higher or lower than standard height?	☐	☐	☐
5	In older buildings, have all steps that could pose a trip or fall risk been identified? (Ensure there are handrails in place on all stairways.)	☐	☐	☐
6		☐	☐	☐
7		☐	☐	☐
8		☐	☐	☐
9		☐	☐	☐
10		☐	☐	☐

Owner:

Assessment date:
Re-assessment date:

Topic	Hazard	Risk	People at Risk	Risk Rating	Control Measures	Actioned X	Residual Risk
Stock room	Falling objects Slips, trips and falls	Personal injury			• **Racking secured to walls or floor**	☐	
					• **Safe system of work for removal of stock from shelving eg kick steps or small steps**	☐	
					• **Stock stacked safely - heavy items at waist height where manual handling will be used**	☐	
					• **Stock on shelves not on floor**	☐	
					• **Daily removal of rubbish**	☐	
						☐	
						☐	
						☐	
						☐	
						☐	

Owner:

Assessment date:
Re-assessment date:

Questions the Risk Assessor needs to ask

No	Question	Yes	No	N/A
1	Is the racking/shelving securely fixed to either wall or floor?	☐	☐	☐
2	Are suitable steps supplied for the removal of stock from shelves?	☐	☐	☐
3	Is there enough space for the safe storage of merchandise (eg can it be stacked without falling off the shelf)?	☐	☐	☐
4	Are all walkways and aisles kept free of stock to prevent slips, trips and falls?	☐	☐	☐
5	Is all rubbish including empty boxes removed regularly?	☐	☐	☐
6		☐	☐	☐
7		☐	☐	☐
8		☐	☐	☐
9		☐	☐	☐
10		☐	☐	☐

Owner:

Assessment date:
Re-assessment date:

Topic	Hazard	Risk	People at Risk	Risk Rating	Control Measures	Actioned X	Residual Risk
Stock room/ despatch	Manual handling	Personal injury, trips and falls			• **Use of mechanical aids where possible**	☐	
					• **Employees trained in manual handling operations**	☐	
					• **Manual handling guidelines observed**	☐	
					• **Team lifting or dismantle load**	☐	
					• **Route clear and safe place to set load down**	☐	
						☐	
						☐	
						☐	
						☐	
						☐	

Owner:

Assessment date:
Re-assessment date:

Questions the Risk Assessor needs to ask

No	Question	Yes	No	N/A
1	Can manual handling operations be avoided by the use of mechanical aids (such as a sack barrow, trolley or pallet truck)?	☐	☐	☐
2	Have all employees who may be involved in manual handling operations been adequately trained to enable safe lifting?	☐	☐	☐
3	Do all employees observe the guidance for the lifting and movement of loads? (Refer to Manual Handling Operations 1992 Guidance)	☐	☐	☐
4	When mechanical aids are not available do employees obtain help from colleagues or reduce the load?	☐	☐	☐
5	When lifting and carrying do employees ensure their route and setting down point is clear?	☐	☐	☐
6		☐	☐	☐
7		☐	☐	☐
8		☐	☐	☐
9		☐	☐	☐
10		☐	☐	☐

Owner:

Assessment date:
Re-assessment date:

Topic	Hazard	Risk	People at Risk	Risk Rating	Control Measures	Actioned X	Residual Risk
Storage of liquid petroleum gas (LPG), calor gas or similar	Gas	Fire Explosion CO poisoning			• **Store minimum quantities of LPG on premises** • **Store LPG cylinders not being used in a steel cage in open air** • **Ensure warning signs are in place to highlight flammable material** • **Ensure storage area locked at all times** • **Ensure different gases, eg LPG, oxygen or acetylene are not stored together**	☐ ☐ ☐ ☐ ☐ ☐ ☐ ☐ ☐ ☐	

Owner:

Assessment date:
Re-assessment date:

Questions the Risk Assessor needs to ask

No	Question	Yes	No	N/A
1	Are the minimum quantities of LPG stored on the premises?	☐	☐	☐
2	Are the LPG cylinders that are not in use stored in a secure steel cage or similar in the open air and separate from flammable material?	☐	☐	☐
3	Are there adequate and suitable warning signs in place to denote fire and explosion hazard?	☐	☐	☐
4	Is there a system that ensures that the secure area for cylinders is locked and that movement of cylinder is recorded?	☐	☐	☐
5	Is there a system to ensure that cylinders containing other gasses eg oxygen or acetylene are not stored with LPG?	☐	☐	☐
6		☐	☐	☐
7		☐	☐	☐
8		☐	☐	☐
9		☐	☐	☐
10		☐	☐	☐

Owner:

Assessment date:
Re-assessment date:

Topic	Hazard	Risk	People at Risk	Risk Rating	Control Measures	Actioned X	Residual Risk
Stress	Stress	Mental illness Other ill-health			• **Training for managers and staff** • **Monitor performance of potential stressors** • **Health and safety and HR work together to combat potential problems** • **Training for managers and staff and repetitive tasks broken up** • **Ensure employees' competence and skills. Provide counselling for employees**	☐ ☐ ☐ ☐ ☐ ☐ ☐ ☐ ☐ ☐	

Owner:

Assessment date:
Re-assessment date:

Questions the Risk Assessor needs to ask

No	Question	Yes	No	N/A
1	Is training provided for managers and staff to enable recognition and management of stressors?	☐	☐	☐
2	Are monitoring and auditing procedures in place to recognise the performance of potential stressors?	☐	☐	☐
3	Does the person responsible for health and safety work together with human resources personnel to combat potential or existing problems?	☐	☐	☐
4	Is there a system in place to ensure that employees are competent for the tasks required of them?	☐	☐	☐
5	Do you provide independent counselling for employees that feel they need it?	☐	☐	☐
6		☐	☐	☐
7		☐	☐	☐
8		☐	☐	☐
9		☐	☐	☐
10		☐	☐	☐

Owner:

Assessment date:
Re-assessment date:

Topic	Hazard	Risk	People at Risk	Risk Rating	Control Measures	Actioned X	Residual Risk
Tempera-ture and ventilation	Tempera-ture Ventilation	Stress Occupa-tional health			• **Ventilation system to be maintained to provide fresh air at a comfortable temperature** • **Keep the temperature constant during opening hours. Ideal for sedentary work is between 20oC and 23oC, must not fall below 16oC. The temperature must have reached 16oC or more after the first hour** • **Thermometers must be displayed for staff use** • **Where possible reduce draughts** • **If the heating, ventilation and air conditioning system has a cooling tower ensure that they are registered with the EHO** • **Regularly clean air ducts and grills and keep free of dust and debris**	☐ ☐ ☐ ☐ ☐ ☐ ☐ ☐ ☐ ☐	

Owner:

Assessment date:
Re-assessment date:

Questions the Risk Assessor needs to ask

No	Question	Yes	No	N/A
1	Do you have a written maintenance schedule for your heating/ventilation system?	☐	☐	☐
2	Are employees given access to HVAC controls to enable adjustment?	☐	☐	☐
3	Are thermometers displayed where employees can monitor temperatures?	☐	☐	☐
4	Have you registered your cooling tower with the local authority Environmental Health Department?	☐	☐	☐
5	Does your maintenance schedule include cleaning ducts and grills?	☐	☐	☐
6		☐	☐	☐
7		☐	☐	☐
8		☐	☐	☐
9		☐	☐	☐
10		☐	☐	☐

Owner:

Assessment date:
Re-assessment date:

Topic	Hazard	Risk	People at Risk	Risk Rating	Control Measures	Actioned X	Residual Risk
Toilets and washing facilities	Poor hygiene	Personal health			• **Safe system of work in place for regular cleaning of facilities** • **Separate facilities provided for males and females, except where each convenience is in a separate room, the door of which is capable of being secured from the inside** • **Suitable sanitary dispensers provided in female toilets, emptied in accordance with waste disposal legislation** • **Hot and cold running water available. Soap and suitable drying facilities available**	☐ ☐ ☐ ☐ ☐ ☐ ☐ ☐ ☐ ☐	

Owner:

Assessment date:
Re-assessment date:

Questions the Risk Assessor needs to ask

No	Question	Yes	No	N/A
1	Are there systems of work in place for regular cleaning of toilet and washing facilities?	☐	☐	☐
2	Are separate toilet facilities provided for male and females? (Refer to the Workplace (Health, Safety and Welfare) Regulations 1992 (SI 1992 No 3004) for ratio of toilets to staff.)	☐	☐	☐
3	Do you have a contract with a feminine hygiene company for the removal and replacement of sanitary dispensers?	☐	☐	☐
4	Are there procedures for ensuring the correct temperature of hot water, and identification of drinking water?	☐	☐	☐
5	Is there a system in place for ensuring soap and drying facilities are maintained?	☐	☐	☐
6		☐	☐	☐
7		☐	☐	☐
8		☐	☐	☐
9		☐	☐	☐
10		☐	☐	☐

Owner:

Assessment date:
Re-assessment date:

Topic	Hazard	Risk	People at Risk	Risk Rating	Control Measures	Actioned X	Residual Risk
Transport (Vans and cars)	Driving accidents	Fatal Serious injury			• **Vehicles used for work must be suitable and fit for the purpose** • **Vehicles must have current road tax, insurance and MOT (if applicable)** • **Procedures in place to ensure vehicles are in good running order with daily/weekly/monthly checks** • **Procedures in place to ensure that all drivers have an appropriate driving licence and are authorised to drive company vehicles** • **Employees instructed to ensure vehicles are not overloaded and that loads are safely distributed**	☐ ☐ ☐ ☐ ☐ ☐ ☐ ☐ ☐ ☐	

Owner:

Assessment date:
Re-assessment date:

Questions the Risk Assessor needs to ask

No	Question	Yes	No	N/A
1	Are all vehicles being used for workplace activities in good condition and fit for the purpose?	☐	☐	☐
2	Is there a system in place to ensure that all vehicles have current road tax, insurance and MOT (if applicable)?	☐	☐	☐
3	Is there a system in place for ensuring that drivers carry out regular checks? Eg daily, weekly, monthly check of oil, water, tyres, washers, lights etc.	☐	☐	☐
4	Is there a system in place to ensure that all drivers have a current driving licence?	☐	☐	☐
5	Is there a company policy to ensure that all vehicles are not overloaded and that loads are safe, secure and well distributed?	☐	☐	☐
6		☐	☐	☐
7		☐	☐	☐
8		☐	☐	☐
9		☐	☐	☐
10		☐	☐	☐

Owner:

Assessment date:
Re-assessment date:

Topic	Hazard	Risk	People at Risk	Risk Rating	Control Measures	Actioned X	Residual Risk
Trolleys	Misuse of trolleys	Cuts			• **Ensure that trolleys are in good condition**	☐	
		Bruises			• **Ensure that aisles are adequate for trolley movement**	☐	
		Falls			• **Ensure that trolleys are collected and located in designated areas**	☐	
		Back injury			• **Ensure signs warn of safe use of trolleys with children**	☐	
					• **Ensure trolleys are of a suitable size and type to meet customer's requirements**	☐	
						☐	
						☐	
						☐	
						☐	
						☐	

Owner:

Assessment date:
Re-assessment date:

Questions the Risk Assessor needs to ask

No	Question	Yes	No	N/A
1	Is there a procedure for ensuring that trolleys are in good working condition?	☐	☐	☐
2	Are aisles of an adequate size to allow free movement of trolleys?	☐	☐	☐
3	Are trolleys kept in specific areas with a system that allows collection after use and re-location to the designated areas?	☐	☐	☐
4	Some trolleys are designed for the transport of children, therefore, is there a system to warn the public of misuse?	☐	☐	☐
5	Are your trolleys of the type that are suitable for the merchandise being sold and which allow customers to use them safely?	☐	☐	☐
6		☐	☐	☐
7		☐	☐	☐
8		☐	☐	☐
9		☐	☐	☐
10		☐	☐	☐

Owner:

Assessment date:
Re-assessment date:

Topic	Hazard	Risk	People at Risk	Risk Rating	Control Measures	Actioned X	Residual Risk
Waste skips for waste food	Manual handling Slips trips and falls Food poisoning	Personal injury/ illness			• **Pest control programme in place. All waste put in skip and surrounding area kept clean** • **Skips in secure locked area** • **Warning signs displayed** • **Skips kept closed when not in use**	☐ ☐ ☐ ☐ ☐ ☐ ☐ ☐ ☐ ☐	

Owner:

Assessment date:
Re-assessment date:

Questions the Risk Assessor needs to ask

No	Question	Yes	No	N/A
1	Do you have an effective pest control programme in place?	☐	☐	☐
2	Are there procedures in place for ensuring that waste is put inside skip and surrounding areas are kept clean and free from debris?	☐	☐	☐
3	Are skips for out of date food kept in a secure area to prevent access by trespassers? (If trespassers get food poisoning from out of date food taken from your skip your company could be liable.)	☐	☐	☐
4	Are hazard-warning signs displayed to deter trespassers?	☐	☐	☐
5	Are employees trained in procedures for keeping skips closed when not in use?	☐	☐	☐
6		☐	☐	☐
7		☐	☐	☐
8		☐	☐	☐
9		☐	☐	☐
10		☐	☐	☐

Owner:

Assessment date:
Re-assessment date:

Topic	Hazard	Risk	People at Risk	Risk Rating	Control Measures	Actioned X	Residual Risk
Waste skips (general)	Commercial waste Manual handling Fire	Occupa-tional health Cuts/ personal injury Slips trips falls			• **Employees, contractors, public not allowed to use skips for disposal of personal waste** • **Personal protective equipment issued to employees** • **Procedures for disposal of different types of waste in place** • **Manual handling training given to employees** • **Copy of waste contractor's registration certificate obtained and waste transfer notes completed**	☐ ☐ ☐ ☐ ☐ ☐ ☐ ☐ ☐ ☐	

Owner:

Assessment date:
Re-assessment date:

Questions the Risk Assessor needs to ask

No	Question	Yes	No	N/A
1	Are procedures in place to prevent employees, contractors and public from tipping their own waste?	☐	☐	☐
2	Has a PPE assessment been carried out to identify the need for PPE?	☐	☐	☐
3	Do you use a registered waste disposal contractor to collect waste? Do you have a copy of their registration certificate?	☐	☐	☐
4	Are there safe systems of work and procedures in place for disposal of different types of waste ie waste oil, glass, and flammable materials?	☐	☐	☐
5	Are all employees involved in waste disposal trained in manual handling operations?	☐	☐	☐
6		☐	☐	☐
7		☐	☐	☐
8		☐	☐	☐
9		☐	☐	☐
10		☐	☐	☐

Owner:

Assessment date:
Re-assessment date:

Topic	Hazard	Risk	People at Risk	Risk Rating	Control Measures	Actioned X	Residual Risk
Windows	Glass	Cuts/ personal injury			• **Ensure that all windows are in good condition and correctly fitted**	☐	
					• **Cracked windows to be replaced**	☐	
					• **System to make safe the area surrounding broken window**	☐	
					• **Ensure open and closing devices on windows and are in good condition**	☐	
					• **Ensure windows can be secured to eliminate falling from/through**	☐	
						☐	
						☐	
						☐	
						☐	
						☐	

Owner:

Assessment date:
Re-assessment date:

Questions the Risk Assessor needs to ask

No	Question	Yes	No	N/A
1	Is there a system in place to check that windows are in good condition and fitted correctly?	☐	☐	☐
2	Does the organisation have a policy and/or arrangements to replace broken or cracked windows?	☐	☐	☐
3	Are broken windows dealt with immediately?	☐	☐	☐
4	Do windows that open and close do so easily and are they in good condition?	☐	☐	☐
5	Can windows be secured so as to eliminate possible falls from or through open windows?	☐	☐	☐
6		☐	☐	☐
7		☐	☐	☐
8		☐	☐	☐
9		☐	☐	☐
10		☐	☐	☐

Owner:

Assessment date:
Re-assessment date:

Topic	Hazard	Risk	People at Risk	Risk Rating	Control Measures	Actioned X	Residual Risk
Work equipment	Circular saw	Cuts			• **Ensure all workplace equipment is suitable and fit for purpose**	☐	
	Meat slicer	Amputation			• **Ensure all equipment has guards and that they are fitted correctly**	☐	
	Etc				• **Ensure that public do not have access to equipment**	☐	
					• **Ensure that staff are competent to use the equipment**	☐	
					• **Ensure equipment is regularly maintained and cleaned and record kept**	☐	
						☐	
						☐	
						☐	
						☐	
						☐	

Owner:

Assessment date:
Re-assessment date:

Questions the Risk Assessor needs to ask

No	Question	Yes	No	N/A
1	Is there a procedure to determine and ensure that all workplace equipment is suitable and fit for purpose?	☐	☐	☐
2	Are there procedures to ensure that guards are correctly fitted and that interlocks, where fitted, are working properly?	☐	☐	☐
3	Is there a system to ensure that members of the public and staff not using equipment do not have access to the work area?	☐	☐	☐
4	Is there a system to ensure that all staff using equipment are competent?	☐	☐	☐
5	Is there a system to ensure that all workplace equipment is regularly maintained and records are kept?	☐	☐	☐
6		☐	☐	☐
7		☐	☐	☐
8		☐	☐	☐
9		☐	☐	☐
10		☐	☐	☐

Owner:

Assessment date:
Re-assessment date:

Topic	Hazard	Risk	People at Risk	Risk Rating	Control Measures	Actioned X	Residual Risk
Young people	Lack of knowledge/ experience	Personal injury			• **Tasks identified prior to starting work**	☐	
					• **Health and safety introduction provided**	☐	
					• **Capabilities identified**	☐	
					• **Training given in all tasks**	☐	
					• **Supervised**	☐	
						☐	
						☐	
						☐	
						☐	
						☐	

Owner:

Assessment date:
Re-assessment date:

Questions the Risk Assessor needs to ask

No	Question	Yes	No	N/A
1	Has risk assessment been completed prior to the young person starting work?	☐	☐	☐
2	Has induction been carried out for basic health and safety?	☐	☐	☐
3	Have the young person's capabilities been identified?	☐	☐	☐
4	Is the young person receiving adequate training for the tasks to be undertaken?	☐	☐	☐
5	Is adequate supervision provided?	☐	☐	☐
6		☐	☐	☐
7		☐	☐	☐
8		☐	☐	☐
9		☐	☐	☐
10		☐	☐	☐

Owner:

Assessment date:
Re-assessment date:

ACTION PLAN

SECTION:

SITE:

RISK ASSESSMENT ITEM NO	ACTION REQUIRED	ACTION BY	PRIORITY 1-5	TARGET DATE	COST	COMPLETED (Name/Signature)
-			-			
-			-			
-			-			
-			-			
-			-			
-			-			
-			-			
-			-			

KEY
Priority
1
2
3
4
5

Action Plan Prepared by (name):

Signature:

Date:

ACTION PLAN

SECTION:

SITE:

RISK ASSESSMENT ITEM NO	ACTION REQUIRED	ACTION BY	PRIORITY 1-5	TARGET DATE	COST	COMPLETED (Name/Signature)
-			-			
-			-			
-			-			
-			-			
-			-			
-			-			
-			-			
-			-			

KEY
Priority
1
2
3
4
5

Action Plan Prepared by (name):

Signature:

Date:

ACTION PLAN

SECTION:

SITE:

RISK ASSESSMENT ITEM NO	ACTION REQUIRED	ACTION BY	PRIORITY 1-5	TARGET DATE	COST	COMPLETED (Name/Signature)
-			-			
-			-			
-			-			
-			-			
-			-			
-			-			
-			-			
-			-			

KEY
Priority
1
2
3
4
5

Action Plan Prepared by (name):

Signature:

Date:

ACTION PLAN

SECTION:

SITE:

RISK ASSESSMENT ITEM NO	ACTION REQUIRED	ACTION BY	PRIORITY 1-5	TARGET DATE	COST	COMPLETED (Name/Signature)
-			-			
-			-			
-			-			
-			-			
-			-			
-			-			
-			-			
-			-			

KEY
Priority
1
2
3
4
5

Action Plan Prepared by (name):

Signature:

Date: